A SPACE FOR SCIENCE

A SPACE FOR SCIENCE

THE DEVELOPMENT
OF THE SCIENTIFIC COMMUNITY IN BRAZIL

SIMON SCHWARTZMAN

The Pennsylvania State University Press
University Park, Pennsylvania

Library of Congress Cataloging-in-Publication Data

Schwartzman, Simon, 1939–
  [Formação da comunidade científica no Brasil. English]
  A space for science : the development of the scientific community in Brazil / Simon Schwartzman.
      p.   cm.
  Rev. translation of: Formação da comunidade científica no Brasil.
  Includes bibliographical references and index.
  ISBN 0-271-00740-0
  1. Science—Brazil—History. I. Title.
  Q127.B8S3913 1991
  509.81—dc20   90-49321
                                                                  CIP

Copyright © 1991 The Pennsylvania State University
All rights reserved
Printed in the United States of America

It is the policy of The Pennsylvania State University Press to use acid-free paper for the first printing of all clothbound books. Publications on uncoated stock satisfy the minimum requirements of American National Standard for Information Sciences—Permanence of Paper for Printed Library Materials, ANSI Z39.48–1984.

CONTENTS

List of Tables	vi
Preface	vii
1. Introduction: A Space for Science	1

PART ONE: FOUNDATIONS

2. The Heritage of the Eighteenth Century	23
3. Imperial Science	47
4. Apex and Crisis of Applied Science	71
5. The 1930 Revolution and the New Universities	105
6. The Roots of Scientific Traditions	139

PART TWO: GROWTH

7. Toward a Scientific Role	167
8. Postwar Modernization	199
9. The Great Leap Forward	215
10. Epilogue	237
Appendix: List of Interviews	249
Bibliographical References	255
Index	269

LIST OF TABLES

1.	Physicists and Geologists, 1892–1907, First Degrees in Brazil	170
2.	Biologists, 1892–1907, First Degrees in Brazil	171
3.	Scientists Educated Abroad, 1892–1907	173
4.	Biologists, 1908–1920, First Degrees in Brazil	176
5.	Physical and Chemical Scientists, 1908–1920, First Degrees in Brazil	180
6.	Physical and Chemical Scientists, 1921–1931	182
7.	Biologists, 1921–1931	184
8.	Rockefeller Foundation Contributions to Science, Research, and Education in Brazil, 1932–1975	194
9.	Structural Changes in Brazilian Society, 1950–1980	200
10.	Growth of the Educational System in Brazil, 1965–1980	220
11.	Students in Graduate Programs, by Field, 1975–1983	224
12.	COPPE: Graduation, Enrollment, Desertion, and Destination of Students, 1965–1978	231

PREFACE

This book started in the mid-1970s at one of Brazil's main science and technology financing agencies, the Financiadora de Estudos e Projetos (FINEP), as a research project aimed at drawing a broad picture of the arrival and growth of empirical sciences in Brazil. The work was carried out on two fronts. First, an effort was made to gather and consolidate as much of the published materials on the history of Brazilian science as possible. Second, lengthy open-ended interviews were carried out with a group of about seventy scientists who played significant roles in this history, either scientifically or institutionally. The text of these interviews and the original tapes are now available for consultation at the Centro de Pesquisa e Documentação em História Contemporânea do Brasil (CPDOC) at the Fundação Getúlio Vargas in Rio de Janeiro.[1]

---

1. CPDOC 1984. The interviews were carried out with the help of the Program of Oral History of CPDOC, headed by Aspásia Alcântara de Camargo. The interview with

Earlier products of this effort, published in Portuguese since 1979, constituted what was probably the first attempt to take a comprehensive look at the development of Brazil's scientific community from its historical roots and to provide a coherent view of its struggle to exist. The first (1979) version of this book relied on contributions of Ricardo Guedes Ferreira Pinto, who worked with the history of physics and engineering; Maria Clara Mariani and Márcia Bandeira de Melo, who concentrated on the biomedical sciences; Tjerk Franken, who worked with the history of institutions and produced a detailed chronology of Brazilian science from 1500 to 1945; Nadja V. X. Souza, who worked on earth sciences and chemistry; Antônio Paim, who studied the Portuguese cultural heritage, the role of positivism in the Brazilian scientific outlook, and the creation of the Universidade do Brasil in the 1930s; and José Murilo de Carvalho, who carried out independent research on the history of the Escola de Minas in Ouro Preto.[2] Joseph Ben-David visited the project at an early stage, writing an insightful report on his perceptions of Brazilian science at the time.[3] The project enjoyed the broad support and sympathy of Brazilian scientists and science policymakers, and it would not have been possible without the personal interest and incentive of José Pelúcio Ferreira, then president of FINEP, who stands as a central figure in the present history of Brazilian science. FINEP supported all the research work and the Portuguese-language edition of the book.

The present book started as a project for a straightforward English translation of the 1979 Portuguese text, which proved to be impossible to achieve. As the translation work proceeded, it became clear not only that the original text had to be revised, corrected, and updated, but also that it had been produced for a different public and with a different emphasis than the present text. While the Portuguese edition was aimed at a wide, educated audience of professionals, teachers, scientists, and policymakers who knew a lot about Brazil and very little of the current literature on the social studies of science and technology, the opposite would be true for an English-language text; while the Brazilian reader would be

---

Gleb Wataghin, the founder of modern physics in Brazil, was carried out independently by the physicist Cylon E. Tricot Gonçalves, of the Universidade Estadual de Campinas, São Paulo. Principal transcription and editing of the interviews were done by Marcílio Morais, Beatriz Resende, and Maria Beatriz de Pena Vogel.

2. Schwartzman 1979. Most of these scholars continued with their own independent work, some of which was later organized in a separate volume. See Schwartzman (ed.) 1982; R. G. F. Pinto 1978; J. M. Carvalho 1978; Mariani 1982a and 1982b; Paim 1982; Nunes, Souza & Schwartzman 1982.

3. Ben-David 1976.

interested in the detailed history of institutions that existed and very often disappeared decades ago, or even in the nineteenth century, the international reader would be concerned with its general meaning and direction.

The present book is a compromise between the two extremes. It is based in part on materials used in the 1979 volume, but it also makes use of subsequent texts and a variety of other sources. Everything is placed within a much more cogent and explicit interpretive framework. Over the past ten years, I hope to have evolved in my perceptions of the role science can play in societies such as Brazil, and this book is accordingly more personal and affirmative than the 1979 text. Most of the general sections of the Portuguese edition have been abandoned, and background materials on Brazil's social and economic history have been added. Historical details have now been placed in footnotes, to preserve their value as references for the specialist and to clear the text for everybody else.

I am indebted to Nancy Stepan for her criticism of an early draft and to Herbert S. Klein for detailed comments and extensive suggestions. I hope the book is now more to their liking. Walzi A. Sampaio da Silva helped with his critical reading of several chapters. Part of the translation was carried on by Diane I. Grosklauss, and Helena Araújo Leite de Vasconcelos was helpful checking the accuracy of names and references. The preparation of the first English draft was made possible by a grant from Brazil's Conselho Nacional de Desenvolvimento Científico e Tecnológico. Finally, I am indebted to Ermínio Martins and Richard Whitley for the incentive to prepare an English version of the 1979 text, which is now this book. Part of the writing was done while I was a visiting associate at the Center for Studies in Higher Education, University of California, Berkeley, during the spring of 1987, and the final text was concluded during a period as a visiting professor at the Instituto de Estudos Avançados, Universidade de São Paulo, in 1988, thanks to a grant from the Ford Foundation.

Along with the support and cooperation received throughout these years, I always enjoyed complete freedom—and, accordingly, bear full responsibility—with respect to how the study was conducted and with respect to the ideas and interpretations put forward here.[4] Thus, the

---

4. I am also responsible for the accuracy of quotations from interviews and for translations of Portuguese texts into English. According to standard usage, Brazilian names (with a few exceptions) are spelled according to modern Brazilian orthography; this applies mostly to double consonants, "i" instead of "y," "s" rather than "z" between vowels, "f" instead of "ph," and rules of accentuation. Foreigners who migrate to Brazil usually adopt

mistakes found in the following pages should not be attributed to my colleagues who participated in different parts of the project or to the agencies that supported our work. I hope that because of our common effort, we all have a better understanding of the history and predicaments of Brazil's scientific community, are better equipped to place this knowledge in a broader interpretive framework, and can therefore be more confident about the future.

Brazilian given names and keep their original surnames, and I have also followed this practice. Brazilian and Portuguese institutions are called by their Portuguese names, with an English translation provided the first time they are introduced.

1

INTRODUCTION: A SPACE FOR SCIENCE

Sisyphus

Cursed by the gods, Sisyphus was condemned to carry a large stone uphill, only to watch it roll back down, and start all over again. The legend of Sisyphus is a proper metaphor for the history of modern science in Brazil, where successes have been few and ephemeral but persistence and enthusiasm have always been present. During hundreds of hours of interviews, the individuals who make up this scientific community revealed themselves to be an extremely lucid and critical group, aware of their limitations but proud of their achievements and optimistic about their role. The persistence of Sisyphus stems not from a rosy view of the future but from the conviction that one is on the right path, might someday reach the boundaries of knowledge and make a meaningful contribution to society, or is at least setting the foundation for the work of future generations. When such a conviction exists, failures and frustrations caused by forces and events outside one's control do not seem so important and do not shake the willingness to start again, if needed, if only to reach the same end.

This is one reason why the present study on the development of modern science in Brazil generated so much interest among Brazilian scientists when it was started in the mid-1970s. In a sense, to contribute to such a study was to retrace the paths taken, to relive successful experiences, to enjoy again the feeling of creative work, to lift the stone back on one's shoulders knowing that one had the strength to carry it. Between 1976 and 1978, dozens of Brazilian scientists from many generations (graduates from 1910 through the 1920s, graduates prior to World War II, and those who began to appear during the 1950s) dedicated an average of four to six hours of their time to recounting their experiences. The interviews focused on the scientists' professional lives, on family environment, secondary and university education, initiation into the sciences, educational experience abroad, professional accomplishments, experience within institutions, personal relationships, their successes and failures.[1] Naturally, the interviews wandered into more general themes: the nature of scientific activity; the Brazilian scientific environment; the meaning, importance, and problems of scientific work within Brazil and in the rest of the world. These statements—taped, transcribed, and edited—constitute an extraordinary chronicle of the experience of bringing modern science to a social and cultural environment as yet unaccustomed to it.

Rich in detail, the testimonies are invaluable. They offer us a picture of the different motivations, values, attitudes, and perceptions these scientists share, a picture of what they have found encouraging and what they have found frustrating. Nothing else could provide this sort of information. Scientific knowledge is usually thought of as a collection of concepts, information, and data having an intrinsic value that does not depend on the individuals responsible for producing it. But perhaps the most important conclusion of the present study is precisely the reaffirmation that science is above all a community of well-educated individuals who enthusiastically bring the best of their intelligence and creativity to their task. The results of their work—articles, scientific data, technological applications—are nothing but the tip of an iceberg that cannot sustain itself without its hidden base, the individuals that produce it.

Oral testimonies also have limitations. Memory is selective. The interpretations each individual constructs about his or her life and experi-

---

1. An absolute majority of the older generation of Brazilian scientists, and all but one of our interviewees, were men. Women began to appear in more significant numbers in Brazilian science with the creation of the Universidade de São Paulo in 1934 (though mostly in the social sciences, which are not covered in the present book).

ences are inevitably influenced by such human attitudes as wariness, favoritism, timidity, or pride. These limitations can be reduced when various testimonies concerning the same facts are available and when other sources of information can be checked. To a certain extent, the coexistence of contradictory versions of the same facts does not mean that some scientists are honest while others are not; rather, within this kaleidoscope, each perception is valid from the personal and psychological perspective of the narrator.

The project was received with interest—sometimes mixed with suspicion—for another, more concrete reason. This was a study carried on with the support of a Brazilian governmental agency, and its intention—to listen to the scientists, to seek their point of view and the value of their contribution, not to exclude anyone for ideological or political reasons—was in stark contrast to the hostility and repression manifested by the military authorities against several of Brazil's best-known scientists only a few years before.

The early 1970s are known in Brazil as the years of "the miracle," the quotation marks suggesting the paradox of high economic growth and a national euphoria owing to the renewed conquest of the World Soccer Cup in 1972, but also what were probably the highest levels of political repression that Brazil has ever experienced. In 1964 the military had seized power after a period of political instability and had started a thorough reorganization of the country's economic and political institutions, with the promise (to be postponed for twenty years) of a quick return to civilian rule. Recessive economic policies in the late 1960s had brought inflation down, and the modernization of the state apparatus, combined with the modernization of the domestic market and an influx of foreign capital, brought yearly growth rates of 10 percent and a sense of national confidence that led to the project of raising Brazil to the status of a world power in one or two decades. Given Brazil's population of more than 100 million,[2] a territory of more than 8 million square kilometers, and the largest industrial park in Latin America (concentrated in the São Paulo region), this ambitious project did not seem too absurd.

The dark side of the miracle was not only poverty and social inequality but also political repression. Economic growth was carried out mostly through income concentration at the top of the social pyramid, and studies done a few years later showed that while income had improved for all social groups during the years of the "miracle," inequality had also

---

2. The 1970 census counted 93.1 million inhabitants; the 1980 census, 119 million. The projection for 1990 is about 150 million. See FIBGE 1987: 52.

increased to an extreme degree.³ The military regime was an uneasy alliance of enlightened technocrats, professional soldiers, and militant anticommunists, and by 1968 the balance had tipped from the former to the latter. At the end of that year, all political activity was prohibited and all forms of political freedom were suspended. In the following years, thousands lost their political rights and public jobs—many of the victims were university professors or researchers in government-controlled institutes—while the military crushed attempts at armed insurrection led mostly by students in Brazil's main urban centers.

Given the conservative ideological bent of the military regime, large sectors of Brazil's educated elites assumed that such a regime could only lead the country to economic backwardness and intellectual obscurantism. This was the prevalent view among Brazil's best-known scientists, who had raised their voices against the country's social, economic, and political inequities in the past and who were among the first to lose their jobs and to be forced into exile by the military after 1964. Already in 1968, however, some agencies in the federal government started to provide resources for projects of scientific and technological development; by the mid-1970s it was already clear that, alongside its authoritarian face, the military regime was opening new spaces for science, technology, and higher education.⁴ After 1975, under the presidency of Ernesto Geisel, the balance had tipped again toward enlightened authoritarianism. Economic liberalism began to lose ground to a renewed belief in economic planning and state intervention, and a long-term plan for political liberalization was announced.⁵ This was the context in which we began our interviews.

---

3. Schwartzman 1980.
4. Despite some ideological similarities, the military regime in Brazil during these years was very different from what was experienced at about the same time in Chile, Uruguay, and Argentina. The latter countries had fairly large and educated middle sectors, most of them directly or indirectly dependent on public employment, which were especially hard-hit by their regimes' programs of political authoritarianism and economic liberalism; they lost jobs and opportunities and were thrown in jail or forced into exile in large numbers. In Brazil, the middle sectors were proportionally much smaller, less politicized, and less dependent on public jobs than their counterparts in neighboring countries; in fact, they were among the chief beneficiaries of the state-induced economic boom in the 1970s. In 1972 a comparative survey showed that while Argentina was a large exporter of university-educated people to other Latin American countries, to Europe, and to the United States, Brazil had one of the highest rates of return, despite the political exile forced on many of its best-known scientists and intellectuals. Only in the late 1980s was Brazil showing signs of approaching the patterns of middle-class displacement typical of its Southern Cone neighbors. See Glaser 1978.
5. For an insider's view of the Geisel periods, see Velloso 1986.

The Construction of a Scientific Community

An earlier proposal for this study, which for circumstantial reasons was never carried out, was put forward by a well-known Brazilian economist and intended to show the historical role played by technology in Brazil's economic development. Besides its eventual academic relevance, that project was meant to provide legitimacy for building up the country's scientific and technological capabilities and, by implication, to bolster the project's sponsor, the Financiadora de Estudos e Projetos, or FINEP (Financing Agency for Studies and Projects), an outgrowth of the Banco Nacional de Desenvolvimento Econômico (National Bank for Economic Development), Brazil's main investment bank. We were able, instead, to move beyond the economist's frame of mind and to try to demonstrate that there was in Brazil a scientific community that predated the rediscovery of economic rationality by the new government agencies, a community that could not be placed under the narrow limits and controls of economic planning and that required freedom of research, permanent public support, and self-regulation as conditions of survival, reproduction, and growth. Against the prevailing economicist mood, we stressed the tensions between science and technology, rather than their complementarity; the links between science and culture and higher education, rather than links with the economy; and the reliance of science on self-regulated groups and institutions, instead of its dependency on the state.[6] The words "scientific community" in the title of the 1979 book (see Preface above) was the last stage in the construction of our research object.[7] As the project developed, FINEP also moved gradually from support of technology to support of science, technology, and graduate education in general.

This choice of subject and approach was based on the belief that science, if understood in very broad terms as a quest for the development of intellectual competence and the enlargement of the pool of knowledge, could play a central role in a country like Brazil, which still faces the problem of how to participate fully in the modern world. Our concern was less with scientific knowledge as such, or with practical applications, than with its role in the broader process of societal rational-

---

6. This concern was not foreign to the one that moved Robert K. Merton to stress the prerequisites of autonomy and self-regulation in his early reflections on science in Nazi Germany. See Merton 1938.

7. This is a post factum reconstruction of a much more erratic and tentative process. The convincing power of the entire enterprise depends very much on the reader. On the construction of research objects, see Latour and Woolgar 1979 and Knorr-Cetina 1981.

ization.[8] How can this kind of knowledge penetrate societies that have not participated in or that have remained at the margins of Europe's scientific revolution since the Renaissance? How does it relate to local values, institutions, and social groups? How is it appropriated by different sectors? How does it take root or remain rootless? Does it really play the role it is supposed to?

Questions like these are broader and hazier than those that hold the attention of most sociologists and historians of science in Western Europe and the United States. In those regions, science is usually thought of as taking place in dynamic scientific centers where great works are written, great discoveries are made, and great theories proposed. The broader context is usually taken for granted. One could argue, following Thomas Kuhn, that such spectacular achievements are only the more visible aspects of everyday scientific activity. A study limited to the great feats of science would suffer the same deficiencies as traditional historiography, starring only kings, popes, and battles. Remarkable people and events do not bring us in touch with everyday reality, and without the latter the very existence of such people and events is not comprehensible. That is what makes modern historiography more social-, economic-, and institution-minded than before. It is for this reason too that one can study the historical and social dimensions of scientific work in regions that are peripheral to the more dynamic scientific centers. This is a study of "normal" science—in fact, the only science Brazilians could have.

Yet this sociology of "normal" science, however needed, could probably be carried out better in other places. The justification for the present study rests on other grounds. First, there was the short-term political motivation of stressing the role and importance of the scientific community in opposition to the technocratic mood that was replacing the obscurantism of the previous years. Less circumstantial is the fact that Brazil is one of only a few countries "south of the equator" that have been able to develop resilient and fairly significant scientific groups and institutions in the twentieth century (the main example, much better analyzed, being India).

To be "south of the equator" means not to have participated fully in the Western intellectual and cultural tradition to which modern science and its related institutions, such as modern universities and entrepreneu-

---

8. This concern with rationalization, inspired by the sociology of Max Weber, should not be mistaken for naive rationalism or for the evolutionist belief that societies are destined to move from lesser to greater rationalization in a process of continuous replacement of all the old, "traditional" forms of knowledge and social organization by "modern" ones. For a contemporary discussion, see Bendix 1984.

rial capitalism, belong. And yet to be peripheral to the Western tradition can mean very different things for different societies. Brazil is a product of a special brand of European civilization, that of the Iberian Peninsula, which did not find in its newly discovered territories a suitable native population and culture upon which to establish its domination.[9] Brazil's colonization was carried out by Portuguese settlers of many different kinds (nobles and courtiers endowed with royal privileges and monopolies, bandits, adventurers and gold seekers, Jesuit missionaries, escaped navy conscripts, new Christians escaping the Inquisition), at first with the help of enslaved Indians, later with African slave labor, and since the late nineteenth century with waves of immigrants from Italy, Germany, various Central European countries, and Japan. The result was one of the largest and more heterogeneous countries in the world, with a current population of about 150 million, a highly industrialized region around São Paulo, areas of intense poverty in the Northeast, European-like regions in Paraná and Santa Catarina, some universities of fairly good quality, and a large number of illiterates.

How does modern science take root and flourish outside its traditional cradle? How does it relate to other intellectual traditions, other institutional settings, other values, other ways of thinking? The growing literature on "peripheral science," which I do not attempt to review here, has gone from diffusionist to imperialist explanations, from analyses of cultural incompatibilities to the search for functional equivalencies, from theories and proposals for scientific and technological modernization to the proclamation of alternative, unique, and supposedly more promising scientific traditions.[10]

Let us deal with these and other questions from what I hope is a more illuminating point of view. Let us take scientific work as one among many

---

9. The Portuguese colonizers found the new territories inhabited by native populations that did not have the same degree of social organization and demographic density the Spanish conquistadores met in Mexico and in the Pacific. As in the United States and Australia, the native peoples were gradually decimated or expelled from the coastal areas to the interior and have remained marginal to the dominant society. The only significant effort to colonize the South Atlantic Indians was carried out by the Jesuits in southern Brazil and was later confined to Paraguay, the only country where Guaraní is still widely spoken and whose population descends predominantly from South Atlantic Indians.

10. For a broad view of the social-scientific literature on higher education and science in Latin America, see Vessuri 1986 and 1987. See also Basalla 1967 for diffusionism; McLeod 1975 and Pyenson 1982 and 1984 for imperialism; Herrera 1971 and Sábato (ed.) 1975 for dependency; Sagasti 1983 for modernization; and Bella 1971 for functional equivalents. The literature on alternative cultural traditions is quite poor in Latin America but very extensive in other Third World regions.

human undertakings, as an instance of human agency that builds and changes social structures in its interactions with other social agents within the boundaries of their time-space constraints.[11] What may be unique in the social study of science in peripheral or semiperipheral societies such as Brazil is the effort to understand how the carriers of modern scientific institutions and culture have had to thread a difficult path between two polar ways of conceiving, organizing, and interpreting what they were trying to accomplish. On one hand are the pragmatists, able to understand, justify, and explain science only through its economic and technological effects; on the other are those for whom science is equated with the free pursuit of knowledge, a noble activity of cultured people.[12] In tracing this path, there will be much to be said and pondered concerning the efforts to establish a space for "normal" science, a modern university system, and an effective way of participating (even if not centrally) on the front lines of scientific activity.

To understand better how the Brazilian scientific community was formed and why it never reached the numerical and qualitative levels attained in other Western countries, we decided to combine the testimonies yielded by our interviews with a survey of numerous, hitherto scattered sources and to draw a broad picture of the social and institutional history of Brazil's main scientific traditions. Whenever possible, we concentrated on science rather than on technology, and on scientific institutions rather than on educational or industrial institutions. Nonetheless, we could not ignore the early medical, engineering, and agricultural schools, the applied research institutes, or the more recently established universities, in which most of Brazil's basic science emerged and developed. The testimonies are confined to natural science, that is, to the so-called hard sciences—physics, chemistry, biology, the earth sciences—

---

11. For an extended elaboration of this approach and its implications, see Giddens 1979, chap. 1; and Giddens 1987: 220–21.

12. In his influential *Science in History*, J. D. Bernal warned that the history of science should take us beyond a lifeless view of the evolution of human knowledge, one that would treat history as if it were a simple and progressive construction of the "ideal edifice of truth." "Such history," says Bernal, "can only be written by neglecting the whole social and material components of science and thus reducing it to inspired nonsense." Such "nonsense" also occurs when one takes the opposite point of view and assumes there is a tight one-to-one relationship between certain characteristics of the productive system and scientific activity. Bernal himself was somewhat responsible for the propagation of this idea when he stated, for instance, that "it is these [productive relations], depending as they do on the technical [means of production], that provide the need for changes in these means and thus give rise to science" (Bernal 1971, 1:50). The contemporary view is best expressed in Kuhn 1977.

with little emphasis on any others.¹³ The social sciences were excluded because of the need to limit the project somewhere and because they present a very different reality—not only with respect to the intellectual production involved but also because in Brazil, with few exceptions, the social sciences have never been institutionalized as the natural sciences have.¹⁴

The Quest for Science

Scientific activity cannot develop and be maintained as a sustained endeavor if it does not have a strong component of self-regulation and self-reference. This contention will be tested repeatedly as we follow the rise and fall of scientific and technological institutions in Brazil since the nineteenth century. At least two conditions are necessary for scientists to maintain their peers as their main reference group. First, society must associate science with progress or in some way recognize the value of the scientist's work. It is this recognition that permits scientists to attain social prestige and attract financial support. Second, and paradoxically, the products of the scientists' labors should not be so profitable as to sidetrack them from the task at hand. When scientists assume positions of responsibility for technological undertakings of ample socioeconomic interest or when they assume the hedonistic stance of maximum yield/minimum labor, it means that their concern with personal intellectual development has been pushed into the background, that other reference groups and other values have taken root, and the quality of their scientific work may be in jeopardy.

Our concern with science as the activity of a self-regulated community, rather than as part of a broader process of social and economic change or an attribute of a special professional education, did not necessarily correspond to what Brazilian scientists themselves thought about this matter. The development of a "scientific role" as a distinctive professional niche has been a peculiar part of the West European tradition since the Renaissance.¹⁵ But as we shall see, it was never obvious for

13. Including mathematics, which in Brazil, is historically almost indistinguishable from physics. For an overview, see Höning and Gomide 1979.
14. The history of social sciences in Brazil is the subject of an ongoing project at the Instituto de Estudos Sociais e Políticos de São Paulo (IDESP) under the direction of Sérgio Miceli. See Miceli 1989.
15. Ben-David 1971.

Brazilian medical doctors and engineers involved in research that their identity as scientists should be stressed and differentiated from their professional roles. Brazilian scientists have, more often than not, stressed the usefulness of their work for the state and for their fellow citizens, rather than the distinctiveness of their role.

The amalgamation of professional and scientific roles was part of a broader view prevailing among Brazilian scientists since the nineteenth century which linked science, progress, modernization, and the establishment of science-based professions. In the early 1950s, European migration, the development of industry and commerce, urban growth—all those transformations that had been gaining speed in Brazil since the 1930s—seemed to converge. Fernando de Azevedo, a leading personality in the establishment of educational institutions in the previous years,[16] expressed the prevailing thinking in Brazil's intellectual circles at that time:

> Inasmuch as the development of industry and the discoveries of physics, chemistry, and experimental sciences tend to further the refinement of the moral and political sciences, it would not be too much to hope that such developments would also add to the wealth of knowledge accumulated by the humanities through observation and through experience with the growing use of modern methods. It is thus certain that we have resolutely entered a stage of cultural renewal, which is expanding and diversifying.[17]

Azevedo was well aware of potential difficulties, of the "reservations and wariness—despite our amazement at the fantastic applications of these discoveries—with which we behold the anxiety-provoking problems posed by profound technical and economic transformations occurring in our world because of the breathtaking acceleration of scientific progress."[18] It was not at all clear just how science could spontaneously accompany and contribute to economic development and modernization. Because of its Iberian cultural past and scholastic tradition, Brazilian society could be expected to resist the influx of the new scientific spirit. "The

---

16. Fernando de Azevedo, a sociologist, participated in the organization of the Universidade de São Paulo and is the editor of a collection of articles that constitutes the fullest treatment to date of the development of Brazilian science (see F. de Azevedo [ed.] 1955).
17. F. de Azevedo (ed.) 1955:35.
18. F. de Azevedo (ed.) 1955:10–11.

progress we have made and upon which we need to reflect," says Azevedo, "should not lead us to harbor illusions about the possible occurrence of pauses, whether shorter or longer, or regressions, however transitory, in one sector or another of the vast domain of scientific studies and research. We are all aware of the origins and ramifications of the old concept of culture and the attitudes that have taken root among us . . . that have left behind strong residues and habits which remain despite the deep transformations that have taken place in society."[19] Dilettantism, a lack of team spirit and cooperation, traditionalism in education, excessive concern with short-term gains—all these posed a threat to the steady progress of the scientific spirit, somehow to be overcome.

The new scientific spirit would have to be introduced, therefore, by political mobilization and propaganda. The scientistic ideology,[20] in Brazil as in other Latin American countries, moved gradually from a few isolated research centers and tiny intellectual circles into the educational system and eventually led to tensions between the research-oriented professors and other sectors of society, including the traditional higher education institutions. One may divide the years of scientistic activism into three general phases or periods. The first, which for Brazil occurred in the years prior to World War II, was related to attempts to build new university institutions that could be established around advanced scientific and cultural centers or institutes. The second, typical of the postwar years, included more ambitious attempts to change completely the traditional university structure and to give scientific and technological research a central role in socioeconomic planning. The third, more typical of the late 1960s and 1970s, is characterized by attempts to create isolated and protected niches for scientific research, supported by a renewed belief in the redeeming value of modern science and technology. A fourth period is probably emerging now, marked by an increased awareness of the distinctiveness of scientific work and of its complex interactions with higher education, technology, and the professions. Such an awareness will impose itself through the sheer force of human reflection on experience and through the weight of current predicaments.

19. F. de Azevedo (ed.) 1955:36.
20. This expression refers to the social movement that appeared in England and other European countries around the seventeenth century and that has been given the name "scientism" by historians and sociologists. In Europe, the early propagandists of science, like their counterparts in Latin America during the twentieth century, were concerned with questions of universal education and with extensive projects for scientific and technological research, which they believed would ensure the conquest of nature and the birth of a new civilization (Ben-David 1971:70).

Science, Technology, and the Professions

To make the development of a scientific community the central focus of our study does not require the assumption that science should be organized according to some idealized model, such as the one put forward by Robert K. Merton some years ago.[21] The concept of scientific community should be understood as an "ideal type," in the Weberian sense. It is an intellectual construct that makes explicit existing social values and actions and that helps us understand their consequences, implications, and tensions with other forms of social action. We can follow the emergence of this ideal type, and some of its implications, from at least three different but converging perspectives: one stemming from the sociology of scientific knowledge, another from an analysis of the interplay between science and technology in contemporary societies, and a third from the sociology of professions.

Sociologists of science tell us that "science" is not an unequivocal concept, that it can mean different things for different people. It can be thought of as a pool of knowledge that is developed, accumulated, transformed, and restructured according to the unique dynamics of each field. It can mean not just any knowledge but a special kind of knowledge, with its own rules (generally explicit ones) on how to incorporate new information and new criteria for validating results. It can refer to a special attitude assumed by scientists, which is called "scientific," meaning that one must incorporate new data and be open to new concepts whenever they appear, following the canons considered appropriate in each field.

A "scientific community," in a broad sense,[22] can be seen as a group of individuals who share "scientific" attitudes and values and who relate to one another through their scientific institutions. Individuals sharing a common background of skills, knowledge, and tacit assumptions about a specific field of knowledge are also said to form a scientific community. In such a community, each individual understands his or her specific area of knowledge and something of the adjacent areas. A certain overlap of work and specializations occurs, but no one has an exhaustive and systematic understanding of the entire field. Another element in the characterization of science as a social system is the existence of an authority system that defends the criteria of probity, plausibility, and acceptabil-

---

21. See Merton 1973. For an extended discussion of the concept, see Mulkay 1977.
22. The different meanings the concept can assume in this perspective is best exemplified by Thomas Kuhn's expansion and diversification of his notion of scientific paradigm in the 1970 postface to *The Structure of Scientific Revolutions*. See Kuhn 1970:174–210.

ity of results—criteria that are generally not an explicit feature of scientific method but that are nevertheless an integral and fundamental part of its workings.[23] Some authors go to the extreme of suggesting that these implicit criteria are what "being scientific" is all about.[24]

This scientific community functions ideally, in the words of Michael Polanyi, as a great and complex republic: "The Republic of Science is a Society of Explorers. Such a society strives toward an unknown future, which it believes to be accessible and worth achieving. The scientist-explorer strives toward a hidden reality, for the sake of intellectual satisfaction. As they satisfy themselves, they enlighten all men and are thus helping society to fulfill its obligation toward intellectual self-improvement."[25] Allowing each explorer a maximum of freedom is seen as the best way to promote this exploration, for it would not be possible to use external, extrascientific criteria to decide what is more or less important for science. The scientific community, then, would operate like a broad market that naturally encourages what is more important and leaves aside what is less significant; it would be up to society as a whole to fund the scientific community without trying to influence how the funds are used.

One criticism addressed to this idealized view is that it derives, at best, from an old-fashioned notion of "little science" which separates science and technology completely. But, starting with (or dramatized by) the Manhattan Project, science seems to have taken a leap toward "big science," characterized by large budgets and highly complex research activities involving the coordinated efforts of hundreds or even thousands of individuals. Whenever research attains this level of cost and complexity, the boundaries of science and technology seem to disappear and the scientific "market" as conceived of by Polanyi is replaced by the logic of the true economic market on the one hand and by that of national policies related to large-scale technological projects on the other.[26]

Jean-Jacques Salomon believes that the roots go back further. He argues that modern science has always sought practical results and that the idea of a distinction between pure and applied knowledge is no more than the vestige of an elitist attitude of Aristotelian-scholastic origin, an attitude that serves as a roadblock to the emergence of modern science. Referring to seventeenth-century Europe, Salomon states that no era better illustrates how science is linked to a complete representation of the world: science viewed as contemplation is part and parcel of the develop-

23. Polanyi 1962.
24. E.g., Barnes 1974; Bloor 1976; Latour and Woolgar 1979; Knorr-Cetina 1981.
25. Polanyi 1968:19.
26. Gibbons and Wittrock (eds.) 1985.

ment of a liberal social order, where "technique" belongs to the artisans who carry out "servile" tasks. Technique is perceived as inferior to science, as the artisan is perceived as inferior to the free individual, the scholar.[27]

With the Renaissance, praxis begins to merit greater esteem; experimental research attains greater dignity and scientific knowledge is assigned a role in the achievement of worldly goals. Descartes, advising Cardinal Richelieu, expressed the meaning that science was to assume from then on: "It would be wise for Your Eminence to grant two or three of your millions in order to undertake all experiments needed to discover the specific nature of each body. I have no doubt but that we could thus attain great knowledge, knowledge that would be much more useful to the public than all the victories that might be won at war."[28]

Nonetheless, Descartes' belief in the usefulness of science did not imply that science and technique were seen as the same thing. His recognition of the value of experimental activity may have meant either that speculative knowledge had become more practice-oriented or that the experimental posture had achieved "dignity" and had been incorporated into academic activity.

We know today that even scientific research that is more academic in nature is guided by strategies that are much more complex than an unbiased quest for knowledge.[29] Polanyi's "Republic of Science" describes part of this reality and much of its ideology, as can be seen by the very acceptance that his proposal for the organization of scientific activity has encountered. The tight bonds linking science, practice, and politics correspond to the other part of reality, which is in turn evident in criticisms of and resistance to the market model.

From a narrow point of view, the passage from "little science" to "big science" can be seen as simply a case of the market of the Republic of Science having been restricted by the ceilings imposed on its historical pattern of exponential growth. The ideal of the Republic of Science has a great deal to do with this spirit of wide horizons, of a never-ending incorporation of new people and new ideas, of the stimulation of experimentation within an ever-expanding system. "Big science" seems to correspond to the point at which this growth begins to go too far, laying the ground for the very planning activities that could restrict the market's free operation.[30]

27. Salomon 1970:30.
28. Quoted in Salomon 1970:39. The translation from the French is mine.
29. Knorr-Cetina and Whitley 1981; Latour and Woolgar 1979.
30. Price 1963.

Added to the exponential growth of science and its costs is the no-less-spectacular growth of its practical results. Research on new materials, electronics, and biology have tremendous social, cultural, and economic impact. Within such a context, it is inevitable that society should demand more of scientists and that scientists in turn should feel greater responsibility for the broader implications of the knowledge they develop. This situation puts the scientist in a dilemma. The more general characteristics of the Republic of Science, focused on developing people's talents to the maximum and linked to a reward system based on intellectual merit, are disturbed when criteria of cost, practical applicability, and social utility begin to intervene. This is a particularly acute problem in scientific communities that lie outside the more important centers: alienation from the scientist's broader social context, or even emigration, may be the price for placing maximum priority on the values of the Republic of Science.

It is not surprising to find that, when questioned, Brazilian scientists and researchers claim that their research decisions are based essentially on their academic interest in the subject matter; in fact, though, the decisions they make are strongly influenced by some combination of practical considerations, material and organizational incentives, and the prevailing lines of research within the institutions where they work.[31] This contradiction reflects the scientists' efforts to see that those values which maximize intellectual merit and scientific recognition prevail with respect to the distribution of rewards, prestige, and resources throughout the educational and scientific system in which they live and work. It is also an indication that they are attuned to the practicalities of the "real world."

The tension that exists between what scientists do and what they believe they should do is only one factor (and not the most important one) hampering the operation of a pure "market" logic. Proponents of the market model for science avail themselves of the classic arguments used by economists to criticize monopoly economies: a tendency to inefficiency, the indefinite maintenance of obsolete institutions and organizations, the creation of increasingly complex and cumbersome planning organizations. On the other hand, there are good reasons to justify the quest for precedence, for preferential allocation of funds, and for the maintenance of protectionist schemes: namely, the need to avert a spontaneous concentration of resources and talent; the need to protect still-fragile initiatives that could be absorbed or wiped out by nondifferenti-

---

31. N. S. Oliveira 1975:115.

ated competition; inevitably high social costs; and distortions that arise from allowing a laissez-faire attitude to prevail within an activity that becomes increasingly expensive and dominated by well-organized professional interest groups.

This dilemma is also apparent in the various policies and philosophies of social groups and government bodies connected either directly or indirectly with science, technology, and higher education. It is at the point where these tendencies meet (not always harmoniously) that science develops or is stalled.

The contrasts between science and technology cannot simply be set aside, for they reflect the deeper question of how scientists define their role in society, how they perceive themselves, and how they hope society will treat them. This fact has been clearly perceived by many of those who were interviewed. The biologist Paulo Emílio Vanzolini,[32] for example, stated that "basic and applied zoology vary only in terms of economic interests; if I study the reproductive strategies of a lizard, for example, this is not applied science. If I do the same thing with a fish that has a certain economic importance, it becomes applied science because it will be important in judging how intensively this species of fish can be exploited." He added: "The difference between pure and applied science does not lie merely in the merit of each, nor in the concept itself, but in the type of animal to which it is applied. This, I think, is the basic part." Vanzolini considers himself a basic researcher who sees as one of his tasks the training of applied researchers, arming them with the methodology appropriate for their work.

Among chemists there seems to be a consensus that physical chemistry is the most theoretical specialization in the discipline. Chemists dedicated to studying the characteristics and components of natural products nonetheless also define themselves as basic researchers, inasmuch as they do not seek immediate economic application: "Our work is to identify substances with different chemical structures. Our interest ends here. There need to be pharmacologists, ecologists, agronomists, veterinarians, etc., who care about this work and try to see to what extent the analysis of Brazilian plants is important in explaining each one of their own phenomena" (Otto Gottlieb, interview). Where one draws the line between "basic," "fundamental," "applied," or "theoretical" research depends less on epistemological notions than on the role that scientists aspire to in their society.

---

32. Biographical notes on each of the scientists interviewed are presented in the Appendix. For a larger biography and a summary of the interviews, see CPDOC 1984.

One can see the same dilemmas through the prism of the sociology of professions. It was never obvious to Brazilian medical doctors and engineers that their identity as scientists should be stressed and differentiated from their professional roles. This, by itself, is not peculiar to Brazil. Medicine, with law, has always been a high-status profession, and engineering in Brazil has followed suit in the French tradition. To herald their professions as "scientific," and therefore endowed with an aura of high competence, was one thing; to renounce the prestige (and often high income) of traditional professions was a different matter. In Brazil, as elsewhere, where the biomedical and physical "science" ends and where the medical and engineering "professions" start is more an organizational matter of academic disciplines and professional institutionalization than it is a matter of well-defined epistemological or functional frontiers. Where this frontier lies, however, is important, for there is little doubt that scientific research cannot advance far if it does not become recognized as an independent professional activity endowed with a degree of self-regulation and freedom from short-term pressures and demands that the liberal professions, autonomous as they may be, never enjoy.

Modern science, technology, and the professions tend to develop in parallel (with large areas of intersection) in societies with strong, endogenous industrial growth. The distinction between pure and applied knowledge is often mostly institutional—academic institutions versus centers for technological research, universities versus technical institutes—but the wealth of resources within more-advanced economies and the experience of cross-fertilization between scientific and technological activities make it appear as if the two have a separate but harmonious development. One paradox of underdeveloped countries is that their scientific activities tend to follow international patterns (since it is the developed nations that provide the education and training for their most qualified scientists), while technology lags behind. The better the scientific work done under these conditions, the more it will tend to contribute toward the central body of knowledge being accumulated in each field. And it is in the more-developed nations that there are greater opportunities for practical application of this knowledge. This explains why science as developed on the periphery is sometimes thought of as "alienated," disconnected from each particular nation's practical needs. Because of this perceived "alienation," scientific institutions often find it difficult to justify their work and to obtain from society the resources and freedom of action needed for their work.

An Outline

The foregoing notions provide a useful lead to the way the present book is organized. Part One deals with the historical foundations of the scientific community through the end of World War II. Part Two is more analytical, discussing specific growth patterns and covering the 1930s to the present day. This distinction is not absolute (there are historical as well as analytical materials in both parts), but it does correspond to a clear change in approach, explained in part by the impossibility of following developments in the second half of the twentieth century with the same kind of detailed attention that one could adopt for some fields up to that point.

I shall devote little space to the early explorers who came to Brazil, for although they often left a significant legacy of observations and studies, they had little contact with Brazilian society and left no disciples or institutions.[33] Brazil was then the largest colony of the Portuguese Empire, and in the next chapter I shall examine how Portugal related to the European scientific revolution in the eighteenth century and what sort of intellectual heritage it left to Brazil. For the Portuguese, Brazil was less a colonization project than a large plantation to be explored. During the first two centuries, sugarcane was paramount in the northeastern states; in the eighteenth century, as sugar prices fell drastically on the world market, gold began to be removed in quantity from Minas Gerais.[34]

Chapter 3 deals with the nineteenth century, which saw the end of the gold boom, the arrival of the Portuguese royal family in Rio de Janeiro in 1808, the demise of slavery in 1889, the exile of Brazil's second emperor and member of the Portuguese royal household (Pedro II), and the beginning of the republican period. Around the second half of the nineteenth century, a new agricultural staple—coffee—began to dominate first the central states of Minas Gerais and Rio de Janeiro and then São Paulo, which started its long ascent as the country's economic and demographic hub. This is when the first professional schools are established, the first scientific groups are formed, and imperial sponsorship for these activities is decisive for their success or failure.

Chapter 4 addresses the transition from the nineteenth to the twenti-

---

33. For an overview, see Oberakcker 1960. See also Albertin and Faria 1984 on the Dutch presence in northern Brazil between the sixteenth and the seventeenth century; Chur, Bertels, Komissarov & Licenko 1981 on the Russian explorer G. I. Langsdorff in the nineteenth century; and Ferri 1979/80 for an extensive bibliography.

34. For a broad view of Brazil's colonial heritage, see Holanda 1960b. On Brazil's colonial economy and society, see Simonsen 1962; C. Prado 1967; Furtado 1968; Lang 1979; and Novais 1981.

eth century and the first decades of the republican period. The republican regime was in large part a recognition of São Paulo's new economic and political role, and it coincided with a large influx of European and Japanese migrants who were to alter dramatically Brazil's ethnic composition from São Paulo southward. This was also a period of transition from the old imperial science to a new concern with applied and practical results, and I shall review the apex and crisis of that process. In Brazilian historiography the year 1930 is usually taken as the point at which Brazil entered the modern world. A new, centralized regime comes to power, industrialization becomes a national concern, the first universities are established, and changes in art and literature, which had begun in the 1920s, increase their presence and influence.[35]

Chapter 5 deals with the impact of these changes on Brazil's scientific and educational institutions, with special emphasis on the creation of the first universities. Chapter 6 brings Part One to a close, surveying the roots of the main scientific traditions that were laid down in those years and that still shape a large part of what the Brazilian scientific community is today.

The pace quickens in Part Two. Chapter 7 provides an overview of the different generations of Brazilian scientists in the twentieth century, their professionalization, and the introduction of ingredients of what might be called a modern scientific "ethos," the definition of a "scientific role." Chapter 8 covers the period of the so-called Second Republic, from 1945 to 1964, and the last two chapters bring us up to the present day, surveying the important scientific and technological buildup of the 1970s and the predicaments of the 1980s.

35. See, for the period, among others, Wirth 1970 and Skidmore 1967.

PART ONE

FOUNDATIONS

2
THE HERITAGE OF THE EIGHTEENTH CENTURY

The European Heritage

At the beginning, science as practiced in Brazil was no more than a pale image of European science, as reflected by Portugal. The structures, institutions, and social forces that gave life to science in the Old World were missing, and whatever scientific achievements are found in Brazil in the past must necessarily be related to European, not Brazilian conditions.

Until the nineteenth century, the institutional history of European science can be told as the history of experimental science's gradual conquest of a central position in Western man's culture and worldview. Experimental science developed outside the traditional universities, and it was only in the nineteenth century that an intimate connection between science and university, now taken for granted, took root. A brief outline of this history is necessary in order to see developments in Brazil in their proper perspective.

A landmark of the long process of legitimation and ascendancy of modern science in Europe was Galileo's attitude of defiance, his questioning of the way in which important truths should be established—according to the authority of Aristotle and Ptolemy, seconded by the church, or guided by empirical observations and carried on according to rational procedures.[1] Galileo's case stands as the last attempt by that era's religious and intellectual establishment to subordinate the findings of physical science to dogma and to the products of a priori reasoning. Thereafter, scientific research prospered, congruent with the individualist ethic of that era's burgeoning capitalism and Protestantism. From its most important seedbed, Italy, modern science was transplanted to soil where it would bear more fruit, France and England; and with the appearance of Charles Darwin's theory of evolution in the nineteenth century, it was the biological sciences' turn to confront the religious dogmas of their time.

Science as developed in these nations did not begin in the universities. The venerable, prestigious European universities, such as Oxford, Cambridge, and Paris, were traditional centers for classical studies and for education in law, medicine, and theology; empirical science was relegated to a secondary role. In England the meeting place for scientists was the Royal Society, established in 1660. According to its founders, the Society's original purpose was highly practical, experimental, and technical.[2] This declaration of purpose was not entirely faithful to reality, though. Few of the Society's main figures were inventors of "useful things," and it was the search for a new and original way of knowing the world, embodied in experimental science, that really served as the force behind the movement to support and encourage scientific research. An entirely new view of nature and of the methods by which it should be approached was being forged, in contrast with the traditional culture then prevailing in universities.

Created in 1666 by Jean-Baptiste Colbert, the French Académie des Sciences had the explicit (and highly practical) goal of allowing for the expansion of France's industry and commerce. Unlike the Royal Society, it was not a society of amateurs but an institution of professionals: twenty scholars supported by the government to solve problems brought forth by the royal ministers. The immediate predecessor of the Paris Académie was the Montmor Académie, which brought together such scientists as Pierre de Fermat, Pascal, and Pierre Gassendi, who corresponded with

---

1. See Burtt 1951:70.
2. In the language of the time, "to improve the knowledge of natural things, and all useful Arts, Manufactures, Mechanick practices, Engynes, and Inventions by Experiments (not meddling with Divinity, Moralls, Politicks, Grammar, Rhetorick, or Logick)" (quoted in Mason 1975:259).

Galileo, Descartes, and Hobbes. Initially, the creation of the Académie des Sciences as a practically oriented government institution was an attempt to save the Montmor Académie, confronted as it was with financial difficulties. At that point, as was to happen so often in the following centuries, scientists managed to persuade the government that they could be useful, that the nation needed them; and they received the support they sought.

The academy's success appears to have been inversely proportional to the conviction with which its initial purposes were maintained. Colbert apparently did no more than provide the academy with general guidelines. Camille Letellier Louvois, his successor, gave the academicians practical tasks, such as designing fountains for the royal palaces or inventing games of chance to entertain the court. The academy suffered during this period, but it was again activated and expanded by Jean-Paul Bignon after 1699.

In both England and France, therefore, the appearance of scientific institutions was clearly aimed at the development of practical and applied knowledge, at the service of the elites. In both cases, there was also a group of eminent scientists who struggled against the traditional culture entrenched within the old universities. The science being created, then, was not intended to serve as a neutral tool, free of moral implications, but was accompanied by a worldview that saw science as the best road to a more precise philosophy, a better understanding of people and nature, and a better society. This new Weltanschauung, which students of the period refer to as "scientistic ideology," was not an isolated event but was a part of the social, economic, and political transformation of European society that we now know as the Industrial Revolution.[3]

The high point of seventeenth-century science came with the publication of Sir Isaac Newton's most important work: *Philosophiae Naturalis Principia Mathematica*. The *Principia* synthesizes and caps the entire process of conceptualization and observation under way at least since Galileo and Kepler began to apply modern mathematics to Copernicus' understanding of the universe. The title of the work attests to an intention of Newtonian science that went far beyond simply explaining certain natural phenomena empirically and for utilitarian purposes. What Newton sought—and achieved—was a new understanding of the universe in which reason combines harmoniously with systematic empirical observation. Thanks to the Newtonian synthesis, modern science consolidated its preeminence over the old scholastic culture in its own language and its

---

3. Ben-David 1971; Bernal 1971; Mason 1975; Cardwell 1972; Merton 1970; Gilpin 1968; Crosland (ed.) 1976.

own style, asserting its claim of independence and superiority with respect to applied knowledge. It was not by chance that many perceived an analogy between the preestablished harmony of the Newtonian universe and the ideals of justice and social wealth, to be created in the years to come through individual initiative and the extensive use of empirical knowledge.

Just as it reached its apex, however, English science seemed to lose its impetus. In 1698, Leibniz and John Wallis (by then the only survivor of the old Philosophical College) were asking themselves about the causes behind the decline they noted in scientific research or, as they put it, "the cause of the present languid state of Philosophy."[4] It is possible that Newton's own work, so apparently perfect, had a paralyzing effect on experimental science, much like a great tree whose shadow hinders the growth of vegetation beneath it. Or perhaps the incipient Industrial Revolution was drawing England's best talent toward other activities.

In agriculture; in the textile industry; in the use of coal as fuel; in mining, transportation, and iron and steel production; and above all with the creation of the steam engine, English technology expanded and diversified. This process coincided with the progressive decline of the Royal Society, which gave way to the "nonconformist" institutions beginning to appear in Great Britain's more industrialized centers: Birmingham's Lunar Society, the Manchester Literary and Philosophical Society, the Edinburgh Philosophical Society. Created in 1831, the British Association for the Advancement of Science eventually became the main institution of the British scientific community.

Despite the dynamic energy shown by Scottish scientific research, scholars studying this era seem to agree that France had become the hub of international science by the middle of the eighteenth century. There, in contrast to events in England, the social revolution that accompanied the Industrial Revolution was not to be bloodless. In France there was an official version of science that posed as neutral and technical and that was embodied in the French Académie. There was also an intellectual and cultural movement surrounding science, a burgeoning "scientistic" ideology that would become known historically as the Enlightenment. Published between 1751 and 1777, Diderot and d'Alambert's *Encyclopédie Française* stands as the great work of French science during that age. Compared with similar works from that time, it proved to be highly theoretical and cultural, not technical and applied. Lavoisier was then the central figure of French science, and the influence of such social thinkers as Saint-Simon, Proudhon, and Rousseau bears witness to the political and social

---

4. Quoted in Mason 1975:280.

leanings of the French intellectual and scientific movement. (In contrast, England was then distinguished mainly by the presence of a very important economic school, led by Adam Smith.) The French Revolution takes Lavoisier to the guillotine, partially a result of obscurantism ("The Republic does not need scientists," the official who condemned him reportedly stated) and partially because of his connection with the ancien régime's tax collection system. But French science did not take long to recover and to carve out a preeminent position in the Western world during the Napoleonic Restoration.

Major Themes

With the Newtonian synthesis established, science at the end of the eighteenth century found itself lacking any central problem. On the other hand, there was a model to follow and, besides, the eighteenth century was a time of vast economic expansion, of taming the wilderness in newly discovered worlds, and of the progressive establishment of new technology.

It is worth listing, albeit briefly, the various areas of interest within science at that time. The naturalists stand out, with their concern to describe and (insofar as possible) systematize the objects found in nature—plants, animals, and minerals. Linnaeus pioneers the fitting of these natural objects into a general classification system and is especially successful within the field of botany. Initially developed as a way to organize information, the Linnaean system soon appears on the French intellectual scene, serving as the basis for Buffon's ambitious *Histoire naturelle*, an attempt to classify all phenomena according to rational principles. The continuation of extensive research and of efforts at systematization pave the way for Charles Darwin's theories of evolution, appearing in the nineteenth century and still exerting their influence today.

The observation of natural objects led inevitably to theories about the development of the planet Earth, also taking inspiration from the principles of a preestablished universal harmony. Confronted with the theories of the "catastrophists," who could not help but note existing signs of great upheavals and dramatic events on the face of the Earth, the former conception was defended under the "uniformitarian theory" of Scotland's James Hutton, whose work, made popular by Sir Charles Lyell in the nineteenth century, contributed toward Darwin's evolutionary synthesis. Opposed on one side by conservatism (which found decisive support in the Newtonian idea of celestial harmony) and on the other by geological catastrophism (which endured until recently as a semiclandestine philosophical and theoretical-interpretive stream of thought), evolutionism is perhaps the clearest example of the inextricable links between science, empirical observation, and the mundane views of the material, social, and political world.[5]

Evolutionism contains the idea of a "natural history," within which archeological observations of geological, zoological, and botanical diversity are brought together. The idea of evolution and progress was not repugnant to the German intellectual environment of the day; yet, the philosophy of nature that prevailed took more inspiration from philosophers and poets—Leibniz and Goethe—than from the mechanistic models of Descartes and Newton. This philosophy presupposed the development of the universe from archetypes, primary monads that contained within themselves all principles of life and movement. Not only was this idea the basis of iatrochemistry—which was to develop within Germany in proximity with alchemy—it also prepared the way for morphological research, where the contributions of Lorenz Oken stand out. From Oken on, a mechanical model of the organization of nature was no longer used, having been replaced with a specifically organic model. The study of biological forms was to merge with the empirical analysis of tissues, pathology, anatomy, and physiology, all of which are closely associated with the development of medicine, thus completing the picture for biology.

In the eighteenth century, too, modern chemistry laid its foundations. Lavoisier introduced quantitative research methods within the field, established the concept of elements, and opened the way for the atomic theory of matter, later to be delineated by John Dalton. This was the time of the first studies on heat and energy, immediately applied in the construction of steam engines in England and later fully consolidated by a new branch of physics, thermodynamics, whose roots lie in the works of the Frenchmen J. B. J. Fourier and Sadi Carnot. It was also the time of the first studies concerning electricity and magnetism, when the experimental results obtained by Stephen Gray, Charles F. Dufay, Benjamin Franklin, Luigi Galvani, Alessandro Volta, and others still had not reached the synthesis that would be attempted with Michael Faraday's electromagnetic induction and James Clerk Maxwell's magnetic field theories in the following century.

---

5. See Gould 1977.

The New Universities

The end of the eighteenth century also saw profound transformations in the Western world's main centers of higher education: England, France, and Germany. The last-mentioned was to dominate the nineteenth century and would heavily influence the U.S. higher education system, which would reach its high point during the following century.

Until the nineteenth century, higher education was still based primarily on the classical tradition. Latin, Greek, and the study of logic and philosophy served as preparation for the main professional careers of that time: medicine, law, and the ministry. During the eighteenth century, however, the development of empirical science had begun to show that an education based entirely on the classics was insufficient. Individuals who had obtained their knowledge outside traditional education began to dispute the privileges and professional monopolies claimed by those few who had a classical education.

Already during the eighteenth century some institutions had begun to offer a much more specialized and technical type of education than that which was offered in traditional universities. Among these, the best-known were the Scottish universities (for medicine) and the French École Nationale de Ponts et Chaussées and the Gergsakademie in Freiburg (for engineering). Around the turn of the century, it seemed clear that the cultivated professions catered to by the more traditional universities and distinguished by their prestige were about to disappear, taking with them the whole system of guilds that had prevailed for centuries and that had been bolstered by the ideal of classical education.[6]

This new view of higher education responded to two types of pressure: (1) the need to incorporate new knowledge produced by burgeoning experimental science and (2) the need to do away with the special privileges of the older professions and guilds, making room for new professions, new schools, and new teaching and learning methods, thus substituting one elite for another.

In no nation did this transformation occur more dramatically than in France. There the Revolution at first abolished the old university, replacing it entirely with professional schools.[7] Later, though, a gradual re-

---

6. Ben-David 1977:36.
7. Writes Ben-David: "The new system that began to emerge in 1794 consisted of a series of professional schools for teachers, doctors, and engineers needed by the state. Scientific studies and scientistic philosophy were to inherit the central place that had been occupied by the classics in both secondary and higher education. Eventually, under Napoleon, the scientific orientation was weakened, the emphasis on the new scientific philoso-

sumption of the older forms of education took place, as part of the postrevolutionary Restoration. For in France, as elsewhere in Europe, there were professional and intellectual groups strong enough and sufficiently well organized to force a good deal of their principles and ideologies on society at large and on the new organizational forms of the university system. As much as they may have wanted to establish new forms of teaching, which would separate the technical from the cultivated professions and eliminate the special privileges of professional groups, the rulers of that period could not fight the monopoly of excellence such professional groups exercised almost by definition.[8]

In reality, the *grandes écoles* created under the Napoleonic system to train the main technical cadre for the state were transformed into centers for the training of the new French intellectual elite. Such schools (the École Polytechnique, the École de Mines, the École Normale Supérieure) began to offer a concentrated, high-level education to an elite, while a mass education system was being developed for the rest of the population at a lower level. Under the new system, specialized learning was seen as a form of intellectual enhancement and improvement of the mind, making its students educated citizens of a new type.[9]

In England there was also a trend toward professionalization of education, although never so strong as in France. Traditional English universities (Oxford, Cambridge) held on to the notion that more specialized study was to be valued not as a way of acquiring practical skills, but as a better way of educating the mind as an end in itself. This insistence made it possible for these universities to maintain an ideal of liberal education not directly oriented toward professional careers while recruiting as professors competent scientists and scholars who were specialists and professionals within their particular fields. In this way, the English system held open an option for a more generic kind of learn-

---

phy was completely abolished, and classical learning was restored to its former importance in secondary schooling. But higher education remained identified with specialized education for various professions" (Ben-David 1977:15–16).

8. "Rulers, however, could effectively control the transmission only of specific techniques. They could preempt the esoteric services of clock repairers or gun makers, but they could not control higher learning, which teaches more than techniques, and which provides scope for intellectual virtuosity and originality. . . . Rulers could grant or deny charters to universities and could buy their support, but they could not control them as they could control a workshop in which masters trained apprentices. Higher learning remained a monopoly of the learned class" (Ben-David 1977:35–36).

9. Gilpin 1968.

ing, focused simply on general education. Later this would take on more complete form in the college system that became generalized within the United States.[10]

It was Germany, however, that was to bring scientific research to the nineteenth-century university and become the model that would influence all others. The reform of the German education system (or Prussian, to be more precise) had its beginnings with the creation of the University of Berlin in 1809. The general context seems to have been set by the existence of an "intelligentsia" that developed under the protection of the state, which meant to guide Prussia down the road to modernization while leaving no room for new social groups or for a plurality of economic and political interests.[11] University activity became one of the few means of access and participation open to these intellectuals, who saw the creation of a modernized university as a way of guaranteeing their presence and importance. This led them to resist the complete professionalization of higher education and to work toward maintaining an integrated learning system through a philosophy of naturalist orientation, the *Naturphilosophie*, which had a much more humanistic and romantic component than the positivist philosophy then spreading from France to the rest of Europe. In 1817, under the leadership of Lorenz Oken, the journal *Isis* was founded in Germany, followed in 1822 by the creation of the Deutsche Naturforscher Versammlung, an association of German-speaking scientists and doctors. The latter group would be responsible for unifying the German scientific community, decades before the political unification of Germany was to be achieved, and would also serve as the inspiration for the British Association for the Advancement of Science.[12]

It is this integrated educational system, directed and guided by professors and intellectuals, which for the first time brings about an effective union of teaching and research. This union takes place initially in the teaching of chemistry, pharmacy, and physiology (which by the nineteenth century had already been sufficiently systematized to allow for coherent and integrated teaching) and in the humanities. The existence of several independent universities competing for talented individuals and drawing their prestige from their professor's academic achievements seems also to have been fundamental. Students who wished to become professors had to learn to do research in order to compete in the

10. Ben-David 1971:75–78, 103–6.
11. Rosemberg 1966; Ringer 1969.
12. Mason 1975:578.

professional marketplace; doctors, chemists, and pharmacists, along with future teachers, could now learn how to carry out scientific research as part of their general education.

The idea of a necessary link between teaching and research spread to other countries, despite the obvious difficulties it presents. There is a natural tension between teaching, which transmits what is already known, and research, which searches for what is not known. This tension can be bypassed in some historical moments and eras; but in Germany, as in a few other nations, it led to the creation of a specific system for scientific research, the Kaiser-Wilhelm-Gesellschaft (later to give way to the Max Planck Institutes). When the North American system later incorporated the idea of bringing together teaching and research, it did so with an important innovation: through graduate schools and regular doctoral courses, it recognized research activity as a profession like any other. In the new graduate programs, research was no longer an auxiliary activity within professional learning, nor a simple teaching method used by the professors; rather, it had its own ends and for the first time assumed primacy within the university. In contrast, doctoral degrees in Europe have generally served mostly as a tool for evaluating and accrediting the scholar, commonly as part of his or her career as a professor and not necessarily linked to a specific research activity. It is against this European background that developments in science and higher education in Portugal and in Brazil must be viewed.

Portugal and Modern Science

At first, Portugal played a pioneering role in the transformations that began shaking Europe as of the Renaissance. It would later assume a marginal role, with profound effects on the cultural heritage Brazil was to receive.[13]

The development of navigation, especially during the fifteenth century, played a significant part in laying down the foundation for a new understanding of nature, crowned at the beginning of the eighteenth century by Newton's work. Prior to these fifteenth-century advances in navigation, the inhabitants of the Iberian Peninsula had already taken to the seas in their fights against the Arabs. One result was Portugal's 1415 conquest of Ceuta, guaranteeing safe navigation through the Strait of

---

13. See Sérgio 1972 for an insightful view of Portugal's history.

Gibraltar and closing off the Continent to further Arab migration. In 1418 Pope Martin V gave his blessing to the Portuguese conquests, bestowing on them the characteristics and functions of a crusade, by an edict known as *Sane Charissimus*. During this period, significant progress was made in shipbuilding. In a revolutionary move, Portugal abandoned the use of galleys, replacing them with caravels.[14]

At the close of the fourteenth century, King João I initiated a new Portuguese dynasty, the Avis dynasty, and around 1420 one of his children, Prince Henrique, organized the Sagres School, dedicated to perfecting nautical instruments and ships and to training navigators and sailors. Portugal's future leadership in the conquest of new lands has been attributed at least partially to Prince Henrique's farsighted initiative in bringing specialists of many nationalities to his school.

During the fifteenth century the Portuguese discovered and colonized the Atlantic islands of Madeira and the Azores, explored the western coast of Africa, and discovered a new sea route to the Orient. In 1498, Vasco da Gama's expedition made its way around the Cape of Good Hope to India. Shortly after that, in 1500, Brazil was discovered.

Through navigation, the Portuguese formulated a new geographic view of the world, a view that was in direct conflict with the Mediterranean view of the planet that Ptolemy had developed at the beginning of Christianity.[15] It is often asked whether the Portuguese worried about

---

14. "The heavy and sturdy carracks designed by the Portuguese did not disintegrate anymore in storms on long soujourns at sea; the wood of which they were built and the way they were careened made them stronger than waves and tides.... All sorts of wind directions, instead of slowing the ships down, were turned into allies by a unique combination of lateen and square rigs. This combination allowed a smaller crew to man a larger ship, which made crew members less vulnerable to malnutrition and plagues, and captains less vulnerable to mutinies. The bigger size of the carracks made it possible to embark bigger guns, which in turn rendered more predictable the outcome of all military encounters with the many tiny pirogues of the natives. The size also made it practical to bring back a bigger cargo" (Latour 1987:221).

15. A Portuguese text from the end of the fifteenth century noted: "That which has been written here should be affirmed in spite of that which was stated by the illustrious Ptolemy, who wrote many good things about the division of the world, but who nonetheless failed here. For he divides the world into three parts: the first, populated, located in the middle of the world; and the North he declares as unpopulated owing to excessive cold; and the equator, he also declares as uninhabited, owing to extreme heat. And all of this we found to be the opposite, because the arctic pole, as we have seen, is inhabited, even to the very top, and the line of the equator is also inhabited, by Negroes, where the number of peoples is so great that it is difficult to believe.... And I can truthfully state that I have seen a great deal of the world" (Diogo Gomes, *As relações do descobrimento da Guiné e das ilhas dos Açores, Madeira e Cabo Verde*, quoted in Saraiva 1955, 2:455).

bringing this set of empirical observations together into a synthesis. For Antônio José Saraiva, such an outcome was inevitable:

> As the southbound caravels opened up the Atlantic, navigators went about substituting their traditional empirical heritage point by point, adapting it to the diverse conditions facing them according to a set of rules that were still empirical but that had been developed from new experiences and with the collaboration of the theoretical science of the astronomers. The direct and systematic observation of nature tended to override the simple empiricism of the navigators. Voyages were full of consequences, which must necessarily be considered when studying the evolution of Portuguese culture until the Renaissance.

Saraiva was drawn to this conclusion because "the most striking tendency that took root during Portugal's expansionist development, and that in certain sectors was linked to Portugal, was the active criticism of experience, and this was the criterion of truth." Portuguese thought was heading toward an integration of its new knowledge through a concept that could restore its culture with "the unity and balance that it had lost because of the navigations."[16]

In support of this hypothesis, peninsular culture could boast of the presence of philosophers considered to be in the forefront of modern thought, such as the Jesuit thinkers Pedro da Fonseca and Francisco Suárez, who addressed problems considered "modern" that would later be taken up by Descartes. In his search for a suitable alternative to Aristotle, Suárez exerted great influence during the seventeenth century, especially in the Protestant nations of Central Europe. His works were studied by Leibniz's masters. As for lay culture, there was Francisco Sanches, a Portuguese from Braga who taught in Montpelier and Toulouse. In his book *Quod Nihil Scitur*, first appearing in Lyons in 1581, then republished in Frankfurt (1628) and in Rotterdam (1649), Sanches combats Aristotelianism and calls for a direct examination of natural phenomena, with experimental data being submitted to the scrutiny of critical judgment.

It was not in Portugal, however, that these precursors of modern philosophy—Pedro da Fonseca, Francisco Suárez, and Francisco Sanches—were to find their greatest welcome. Winds were already blowing in another direction.

---

16. Saraiva 1955, vol. 2, chap. 4.

The Counterreform

Around the end of the sixteenth century, Ignatius of Loyola's Society of Jesus, created in 1534, overcame its initial vacillations and opted for the preservation of traditional heritage, as expressed in the Aristotelian-Thomist doctrine. An abrupt reorientation of Portuguese culture was to be fostered by the Jesuit order—averse to contemplation, rigidly hierarchical, militant, devoted, and active. Two tools were used to reach their objectives: the *Ratio Studiorum* and the Inquisition.

The *Ratio Studiorum*[17] synthesized the Jesuits' pedagogical experience and assumed its definitive form at the beginning of the eighteenth century. It laid down the rules for the courses, programs, methods, and disciplines used in the society's schools. Through a set of explicit rules of teaching, it set the norms not only at the so-called lower level but also at the university level. Knowledge was seen as fully systematized; and at the apex of the pyramid stood theology taught according to Thomas Aquinas, followed by philosophy taught according to Thomist Aristotelianism.

The overriding goal was to preserve established knowledge and to prevent any possible epistemological innovation. The Jesuits were not opposed to new information or techniques; but they would not tolerate the broader philosophical views and innovative intellectual institutions that had arisen in some parts of Europe. The questions that teachers were to raise and the texts that students were to read were subject to strict control. Obedience to religious authorities was to be paid in all matters of discipline and study; no reference was to be made to unauthorized books or authors when giving explanations; no new teaching or discussion methods were to be introduced. No one was to be allowed to introduce new questions, nor to present an opinion that was not that of a qualified author, unless duly authorized.

Student access to books was limited to St. Thomas's *Summa Theologica*, Aristotle's philosophical works, selected commentaries, and books that were aimed at a cultivation of the humanities. The Aristotelian doctrine was jealously guarded against any interpretation not approved by church officials, an attitude that contrasted sharply with the openness and flexibility of such luminaries as Suárez in previous years.

This pedagogical doctrine was not used just to preserve the integrity and purity of a single religious order but was made a norm for the entire Portuguese nation. The Jesuits assumed control of education at all levels. In the Universidade de Évora this took place directly; at the Universidade de Coimbra, it was achieved through the Colégio das Artes, which

---

17. Franca 1952.

all students had to attend and where the Jesuits taught the propaedeutic disciplines. In addition, this doctrine permeated the state administration.[18] The result was an impenetrable barrier around Portugal, totally isolating it from modern culture.

The control the Jesuit order exercised over the pedagogical system was aided by the Inquisition. Officially entitled the Tribunal do Santo Ofício, the Inquisition had the responsibility of safeguarding the integrity of the holy Catholic faith. To carry out this task, it was granted broad powers over personal freedom and was even allowed to extract confessions under torture. At the end of the fifteenth century, the Inquisition's activities almost came to a halt in Europe. As part of the Catholic church's struggle against Protestantism, however, those activities were renewed in Portugal in 1540 and were expanded at the beginning of the seventeenth century.

Portuguese historians have been unable to reconstruct the activities of the Inquisition in their entirety. Each case generated a file, and although many have been lost, there remain 36,000 files to be searched. By 1732, according to Saraiva, repenters totaled some 23,068. The number of individuals condemned by the Santo Ofício is estimated to have averaged 120 to 160 per year. Nor were the Inquisition's repressive measures limited to its direct victims. It drove into a state of panic all who were associated with the victims themselves and anyone who aspired to a modicum of free thinking. One of the few existing surveys on the social origin of those condemned between 1682 and 1891 illustrates this well: about 57 percent were from upper or intellectual classes; 30 percent were artisans ("mechanical tradesmen"); and only 12 percent were common laborers. It thus seems reasonable to conclude that the preferred targets were those segments of the population which might oppose the monolithic culture and the cordon sanitaire that protected it from counterinfluences from abroad.[19] Most victims were former Jews, or *cristãos novos* (new Christians), who remained under suspicion and surveillance long after changing their family names to Portuguese denominations of

---

18. One Jesuit priest described the situation thus: "Not anywhere in Europe, or in either of the two hemispheres, was there any nation where our society was more esteemed, more powerful, or more firmly established than in Portugal, and in all of the nations or kingdoms subject to Portuguese rule.... We were more than just guides of the consciences of princes and princesses of the royal family, for the king and his ministers requested our advice on all matters—even the most important—and no position in the government of the state or within the church was filled without first consulting us or without our influence. The high clergy, the powerful, and the people thus fervently fought for our protection and favor" (Anais da Sociedade, quoted in Domingues 1963:109).

19. Saraiva 1955, 2:79–82.

plants and animals, no matter how much they tried to be true to the official faith.

In Portugal, the Inquisition was under the control of the Dominicans; the Jesuits took care of education. Jesuits and Dominicans acted in unison to ensure that their way of thinking would dominate throughout the seventeenth century and the first half of the eighteenth. Although the Dominicans held "great repressive power," in the words of Mário Domingues, the Jesuits held "most of the institutes of learning, where they molded the spirits of rulers; as is well known, they were also the confessors and spiritual guides of the royal family and most of the nobility."[20] During the eighteenth century, notably under King João V, it is believed that a tendency arose within the court to encourage rivalry between the religious orders, in the hopes of reducing their power. The Congregação do Oratório (Congregation of the Oratory) was to play a central role in this process.[21]

It was only at the end of João V's long reign (1706–50) that some light found its way into Portugal through the dark curtain of totalitarian control. This illumination was made possible thanks to certain diplomats who, upon returning to Portugal after having socialized at the Paris and London courts, realized just how backward Portugal had become. At least one of these, Alexandre de Gusmão, rose within the government and was made responsible for some initiatives that were later to bear fruit. Notable among these was granting the Congregation of the Oratory the right to prepare candidates for university entrance. No longer was it mandatory for those who graduated to attend the Colégio das Artes. Thus, the Jesuit monopoly over an essential sector was effectively brought to a close.

With respect to modernization, the most significant event was the publication in 1746 and 1747 of the *Verdadeiro método de estudar* (The True Method of Studying). Its author, Luís Antonio Verney, was a prominent figure in the Congregation of the Oratory.[22] The book comprises a series of letters, published without the signature of their author, who was in Rome. Those letters, addressed to an imaginary interlocutor, provide a

---

20. Domingues 1963:264–65.
21. "The Congregação do Oratório had been founded in Rome in 1550 by Philip Neri and introduced in France by the Cardinal of Berulle in 1611 and in Portugal in 1688 by initiative of Father Bartolomeu do Quental, preacher and confessor at the royal chapel. The Congregation was known in France for its liberalism and for its cultivation of mathematics, physics, the natural sciences, history, and the national language. Malebranche, a disciple of Descartes, was an Oratorian, and the priests of the Congregation always leaned toward Cartesianism" (C. Magalhães 1967:173).
22. Verney 1949/50.

complete and thorough criticism of the Jesuit's pedagogical system. After nearly two centuries of silence and apathy, the Portuguese intellectuals had entered a debate that made them conscious of the need for reform.

Between 1748 and 1756, twenty books and pamphlets were published in defense of or in disagreement with the *Verdadeiro método*. The fiercest opponents demanded an auto-da-fé and destruction of the dangerous text.[23] This time, though, the books were not burned. The reform that was so highly recommended by Verney was to affect all courses taught in Portugal, from Latin and the humanities to technical and professional training. The essence of his message was focused on a radical break with Thomist Aristotelianism. Pointing Portugal down the road to empiricism, Verney argued that authentic philosophy is "knowing what really makes the water rise into a syringe." When Pombal expelled the Jesuits in 1759 and set out to implant a new mentality, he found the ground had already been laid by Verney.

The Pombal Reform

Those who had lived abroad in the service of the king, or for some other reason, and had returned with the intention of rescuing the nation from continued medievalism, of introducing it to modern times, were known in Portugal as *estrangeirados*, or "foreignized." The most illustrious of the estrangeirados was Sebastião José de Carvalho e Melo, later named the Marquis of Pombal. In 1738 he had been appointed ambassador to London, where he lived for several years. After the death of King João V in 1750 and King José I's subsequent ascent to the throne, Sebastião de Carvalho e Melo was invited to join the government, finally becoming the highest authority and de facto ruler. Pombal viewed England's success as stemming from the application of scientific knowledge to productive activities. This was the concept he would try to apply in Portugal.[24]

In 1771, Pombal founded Lisbon's Colégio dos Nobres, a boarding school where a hundred students from the nobility were taught not only

---

23. "Whenever the true author does not come forward, his writings themselves should be made to pay, serving as a statue of the author. Praise the Lord! How long it has been since Portugal has seen one of these bonfires or has offered Christian charity and the public peace the smoke of this holocaust, more precious to it than any incense" (Cândido de Lacerda, in 1749, quoted in J. de Carvalho 1950, 17).

24. See Falcon 1982 for a scholarly account of Pombal's enterprise.

the classics but also mathematics, physics, hydrostatics, hydraulics, drawing, and architecture, all under the strictest discipline and by instructors imported from France and England. The goal was to build a modern nobility that would remain faithful to King José and his powerful minister. But this initiative did not produce the desired results, apparently because of the reigning atmosphere of denunciations and spying. A few years later, Pombal decided to reform the Universidade de Coimbra itself.

The 1772 reform of Coimbra amounted to the founding of an entirely new university. Schools, institutions of practical training, study programs and methods, disciplinary measures and sanctions, buildings, textbooks—all this was at least deeply remodeled and renovated, if not created anew. Most professors were chosen and appointed by Pombal himself, who recruited renowned professors from abroad, especially Italians. Two new schools, Mathematics and Philosophy, were created, the latter being chiefly concerned with what was then known as "natural philosophy," though focused on applied knowledge. Secondary education underwent a total change. The university gained a botanical garden, a laboratory of physics and chemistry, a pharmaceutical dispensary, and a laboratory of anatomy.[25]

Modern scientific thinking had to coexist with autocracy. Pombal wanted

> to civilize the nation while enslaving it, to spread the light of philosophical sciences and to transform royal power into despotism. He stimulated the study of natural, international, and universal law and provided them with chairs at the university; but he did not realize that he was giving lights for the people to see that the government had to serve the nation's, not the prince's well being and had to be bounded in its powers.[26]

25. The relevant statutes sought to implant a new pedagogical style: "Instill the students with the scientific spirit: this point is continually stressed. Instead of useless scholastics, knowledge of the Newtonian laws set out in natural philosophy is prescribed. All theoretical reasoning will be derived from principles fully proven by any of the basic subjects—physics, mathematics, chemistry, botanics, pharmacology, and anatomy." In explaining how a healthy body works, "the professor will describe the part in question, no changes being made by the imagination but rather according to the anatomy: likewise, the movement of fluids will be studied, without hypotheses or fantasies but as shown in experiments, through anatomic injections, animal dissections, everything being explained as far as possible in relation to the laws of physics, mechanics, hydraulics. Medical theory requires caution herein, as well as a strong awareness of its limits. Never should it be insinuated that illness is cured through speculation" (quoted in Cidade 1969, 2:210).

26. Ribeiro dos Santos, quoted in Sérgio 1972:76.

Modern science, in other words, came without its philosophical and ethical dimensions, a feature that could be traced, according to some authors, to the time of the discoveries and that would explain, in the end, Iberia's historical lack of participation in the development of modern science. And so it reached Brazil.[27]

After Pombal's death, a restoration movement known as the *viradeira* (turnabout) destroyed much of what he had accomplished. Antônio Sérgio believed that Pombal's reforms gained substance by the continuous work of the Portuguese Academy of Sciences and through the fellowships that provided for studies abroad in the following years.[28] The French invasions would bring this process to a halt, but at the end of the century Portugal already boasted a significant number of naturalists, mineralogists, metallurgists, and botanists, some well known in the rest of Europe.

State, Church, and Education in Brazil

The previous discussion helps dispel the notion that Brazil was historically a rural, traditional, and deeply Catholic society which then evolved gradually into modernity, a myth that has not resisted modern historiography.[29] As a mercantilist and seaborne empire, Portugal never shared the feudal experience of decentralized power that was dominant in most countries of Western Europe. Its centralized, bureaucratic, and patrimonial administration was transplanted to Brazil—first with the establishment of a general government in 1548 and, much later, with the migration of the whole Portuguese court to Rio in 1808. When Brazil was made independent in 1822 by a member of Portuguese royalty, the line of continuity was never completely broken, and this fact is important for an understanding of the stable institutionalization of the Brazilian government during the colonial period and during the second half of the nineteenth century, in sharp contrast with most of the continent. From this perspective, the republican decentralization of 1889 can be seen as just a pause in a trend that would be taken up again in 1930.

This centralizing tendency explains why, contrary to what is usually thought, Brazil was never a country where the church held undisputed control and authority, even though—or because—the intimate relations

---

27. See M. B. N. Silva 1988.
28. Sérgio 1972:105–8.
29. Faoro 1958; Schwartzman 1973, 1975, and 1982; Velho 1976; E. P. Reis 1979.

that always existed between church and state in Portugal were brought to the Brazilian colony[30] and continued to exist in the Brazilian Empire. It is true, though, that Brazilians would usually declare themselves to be Catholic, and the church did provide the only legitimate ethical and moral code available to the population. The church also had a monopoly over the principal rites of passage that defined one's place in society— baptism, marriage, burial—and to be outside the church meant that one did not enjoy the rights of citizenship such rituals symbolized. More than symbols were involved: throughout the nineteenth century, a Catholic oath was required in order to graduate from the state faculties, to work as a public employee, or to be a member of the legislature.[31] The state was linked to the church through an agreement known as the *padroado,* whereby the state had the right to approve all documents generated by the Roman church before they could be enacted in Brazil; in addition, the civilian authorities participated in the nomination of all Brazilian bishops. This intertwining of church and state meant that, in practice, religious questions were often treated as merely political, and religion was often used to further the political goals of the state. If Brazil were a deeply religious society, this arrangement would have brought into being a theocratic regime; the church hierarchy would have fully controlled both state and society. What happened was almost the opposite: dominance belonged to a secular state, and the church had to play a minor role, accepting without question civil authority and the not-so-Christian mores of the people in exchange for some measure of authority and power.

The consequence of this arrangement was that, for most Brazilians, Catholicism became above all a set of conventional behaviors instead of a deeply felt commitment to religion. More intense forms of religiosity emerged, and still do today, at the bottom of society, independent from and often outside the control of ecclesiastical authority: syncretic cults, millenarian movements and, more recently, Spiritualist and Protestant fundamentalism.

There was also a deep distinction within the church between the religious orders (notably the Jesuits) and the secular clergy working in parishes throughout the country. The Jesuits were hierarchically organized along military lines, and their organization went beyond national frontiers. They controlled most forms of education in the Portuguese Empire until their expulsion in 1759, and they clearly were involved in a project for secular power that extended from doctrinaire control of the Univer-

30. Lacombe 1960.
31. Barros 1962:330.

sidade de Coimbra to the political, economic, and military organization of South American Indians in the Missões region, where the Portuguese and the Spanish empires met. The sheer grandiosity and ambition of this project explains the conflict between the Jesuits and the Portuguese crown and, in the end, the order's expulsion from the Empire.

The secular priesthood was quite another matter. A priestly career was often the only choice for men of obscure social origin who could not expect to attend the Universidade of Coimbra or the faculties established in Rio de Janeiro, São Paulo, Recife, and Salvador in the nineteenth century.[32] Working in small towns or in the countryside, the secular priest's survival depended mostly on the protection and support of the local elite. He would perform the expected rituals and teach religion and the alphabet to the children of the richest families of the region. In the eyes of the political authorities, then, the secular priest not only did not challenge the regime, he contributed to its stability.

Religious education also had two completely different meanings. For the Jesuits, it was perceived as an instrument for controlling society and keeping the civilian authorities under their authority. For the secular priest, it was just a traditional way of raising the children and imbuing them (especially the girls) with Christian virtues. This distinction was perceived clearly by the Portuguese and later the Brazilian authorities. They fought against the Jesuits and when necessary clashed violently with the organized church, but they never ceased to declare themselves Catholic and to bring the children to the church for their education.

When the traditional priests tried to move away from their expected role, they also often moved away from the established church. The best example was probably the intellectual fermentation of the Olinda seminar, headed by Azeredo Coutinho and described as "perhaps the best embodiment of the Brazilian enlightenment—both religious and rational, realist and utopian, mixing the influence of the [French] philosophers with clerical vigilantism."[33] This seemingly incongruous combination of ideas made sense from the peculiar condition of these "liberal priests," who were to play an important role in the movements toward independence in the eighteenth century, who joined the Freemasons, and who even tried to convince the Brazilian Empire to decree the end of celibacy for Brazilian priests, which would be tantamount to the establishment of a national church.

Political independence from Portugal would only strengthen these tendencies. The Empire would keep Catholicism as the official religion, the

32. J. M. Carvalho 1980.
33. Souza 1960:102.

institution of "padroado," and the delegation of civilian rituals to the church. It was, however, a weak church, infiltrated by the Enlightenment and without the strength of the Jesuits, and made weaker by the strong influence of naturalist and scientific ideas among the country's better-educated elites. In the nineteenth century no religious educational institution ever had the prestige and appeal of the professional schools established by the civilian authorities in the main cities. If this subordination of the church could create fermentation among the low clergy, it was never enough to challenge the strength of the civilian state. When this challenge did occur, in an episode that became known as the "religious question" at the end of the century, it was an attempt to reestablish the conservative power of the church's hierarchy rather than something coming from enlightened intellectuals.[34]

We can now understand why the Portuguese never created in Brazil universities like the ones established by Spain in its American colonies. It was too late for Catholic universities in the traditional sense but too early for modern ones.

Projects for a Brazilian University

In colonial Brazil there was no organized higher education; very little teaching beyond elementary classes offered by the church took place at all. As an ally of England, Portugal was invaded during the Napoleonic wars by French troops led by General Junot in 1808. The Portuguese royal family fled to Brazil under protection of the English fleet. Because of the transfer, Brazil was promoted to the "United Kingdom of

---

34. The issue in the "religious question" was whether the Bishop of Olinda, Dom Vital (and later also the Bishop of Belém, Macedo Costa), held the right to expel the members of religious brotherhoods who were also Freemasons or to close these brotherhoods if they resisted their orders. The difficulty was that these brotherhoods were not just religious associations; they also performed several civilian functions. The dispute evolved into a conflict between the relative powers of church and state in a period when the Roman church was trying to reestablish its leadership and authority throughout the world by reaffirming its most traditional and conservative values. Pope Pius IX, in his encyclical *Quanta Cura*, listed all the evils of modern society condemned by the church: rationalism of all kinds; naturalism; indifferentism; the notion of a free church in a free state (i.e., the separation between state and church); the prevalence of civilian authority; the subordination of religious authority to civilian government; liberalism; progress; modern civilization (Barros 1962:349). The Brazilian bishops opposed all these ideas, and their confrontation with the enlightened Empire of Pedro II was unavoidable. For resisting the authority of the civilian state, Dom Vital was sentenced to jail.

Portugal," and Rio de Janeiro became the real capital of the Portuguese Empire.

The Portuguese court brought the colony many innovations. Over the next ten years Brazil would establish courses of higher learning in engineering and medicine and training courses for various professions. The creation of a university was only to be entertained at the end of this period; the initiative is associated with José Bonifácio de Andrada e Silva, who had studied in the reformed Universidade de Coimbra during the last decades of the eighteenth century.

Andrada e Silva came from a well-to-do family of recent Portuguese descent that had made its home in Santos, Brazil. Sent abroad to study at Coimbra in the early 1780s, he completed his course at the Faculdade de Filosofia in 1787 and at the Faculdade de Leis the following year. He preferred a career as a naturalist to teaching, and in 1785 he was made a free member of Lisbon's Academia de Ciências. Already the following year he presented the academy with an essay on whale fishing and oil extraction. In 1790 the Portuguese government sent him on a scientific mission to the rest of Europe, especially to pursue new knowledge in mineralogy.[35]

During part of 1790 and 1791, Andrada e Silva studied chemistry and mineralogy in Paris. In 1792 he left Paris for Freiburg, where he joined the laboratory of Abraham Werner, who is considered the founder of systematic mineralogy. During the following years, he dedicated himself to mineral research in various European nations. These activities earned him admittance to several European scientific institutions, such as London's Geological Society and Jena's Mineralogy Society, and to the corresponding associations in Paris, Berlin, and Edinburgh. He returned to Paris at the end of 1800. Later, Andrada e Silva held important positions within the Portuguese administration. In 1801 he was appointed director of the mining bureau, where he was to be responsible for managing coal mines and reforming the iron processing plants. He also became head of a laboratory for chemical and metallurgical experiments, took over the chair of metallurgy at Coimbra, and remained active in the Portuguese Academia de Ciências, where he served as secretary years later. He returned to Brazil only in 1819.

In the three decades following his graduation from Coimbra, Andrada e Silva remained faithful to the understanding of science that prevailed in Pombal's reform of the Portuguese university, namely that science's

---

35. Falcão (ed.) 1965.

end goal is its application.³⁶ Moreover, success was assured only when scientific knowledge and productive activity interacted.³⁷

It is believed that Andrada e Silva returned to Brazil at the invitation of King João VI to head the Instituto Acadêmico, the type of university that the Portuguese were considering founding in Rio de Janeiro. Appointed in 1821 to write the instructions to the São Paulo representatives to the Portuguese Parliament in Lisbon, he returned to the idea of a Brazilian university, taking much of his inspiration from the Pombal model. No one knows just what led to the postponement of this undertaking, but we do know that a little more than a year after he returned to Brazil to reside in Santos, Andrada e Silva played a central role in the events that led the court to return to Portugal and to the declaration of Brazil's independence in 1822.

The university conceived of by Andrada e Silva would be made up of three schools: philosophy, jurisprudence, and medicine. The school of canons and theology that had been retained under the Pombal reform was omitted, apparently showing that lay thinking had advanced since then. The school of philosophy would be divided into three areas: natural sciences, rational and moral philosophy, and mathematical sciences. This structure, in which mathematics was to be placed within philosophy, was a denial of the autonomy of the school of mathematics granted under the 1772 reform. The teaching of natural sciences would be centered on the study of natural history, chemistry, physics, and mineralogy, the latter examined as thoroughly as possible. Hopeful about Brazil's mineral potential, particularly given its large territory, the course would educate individuals who could take charge of this exploitation.

36. Paim 1971.
37. In 1813, in an essay on coal mines and iron foundries, Bonifácio stated: "If our nation is sterile in agricultural products; if factories face almost insurmountable obstacles in competing with those abroad, what more natural and safe way would a nation have of not becoming impoverished and deserted, if not through the extensive mining of the minerals with which Providence has chosen to bless it? . . . If Russia, Prussia, and France have regained so much wealth through the exploitation of their minerals, who is to keep Portugal from doing the same? Nations are supported and defended by bread, gunpowder, and metals; without these from their own backyard, the existence and liberty of any nation is precarious" (quoted in Falcão [ed.] 1965, 1:40). At the beginning of his career, in his previously mentioned first essay to the Academy, Andrada e Silva wrote: "Common men believe that common things are not part of science; this means that the art of making furnaces is considered vulgar, as is the art of any ignorant bricklayer. Nevertheless, a good deal of knowledge of physics is needed. In Santa Catarina, where Brazil's largest coal mines are located, there are at least twenty boilers with as many furnaces; but if the first builders had known a bit more about the physics and chemistry of fire, all of these could have been reduced to five at most" (Falcão [ed.] 1965, 1:40).

The project never materialized. But even if it had, it most likely would not have succeeded in blending teaching, research, and professional education as in the European universities that underwent modernization in the nineteenth century. The European universities managed in different ways to combine and reconcile the more traditional characteristics of guilds with pressures exerted by burgeoning professional groups bearing the banner and creed of empirical science and the ideals of rationality. In Europe, university autonomy was identified with self-government by a community of scholars and scientists.[38] In the Luso-Brazilian experience, however, the notion of university autonomy tended to be identified with university control by the clergy against a modernizing state. In opposing this autonomy, the Portuguese and Brazilian elites were left with only one of the two key ingredients of modern European universities, namely the education for the professions, but missed the other, their tradition of self-rule and free inquiry.

In short, both Portugal and Brazil lacked a deeper social movement that could look for a renewed university as an instrument of social mobility and affirmation. The transformations that took place were attempts from the top down to create technically skilled individuals to manage state affairs and discover new wealth. As we will see later, this was partially achieved, but there was no space for scientific activities to bear fruit. In assuming an independent path, Brazilian culture incorporated only one component of that day's modern idea of science, the one related to its application. A key component was missing: broad sectors of society that saw in the development of science and the expansion of education the road to its own progress.

38. Rothblatt 1985.

3

Imperial Science

The arrival of political independence in Brazil was mostly smooth and peaceful, thanks to the transfer of the Portuguese court to Rio de Janeiro in 1808 under British protection during the Napoleonic wars. For about twenty years, Rio de Janeiro remained the seat of the Portuguese Empire, and attempts to bring the country back to colonial status in 1822 led to its formal independence under the heir of the Portuguese crown, Pedro I. The first decades of the nineteenth century saw relative economic stagnation, with the exhaustion of the gold mines in Minas Gerais. The expansion of international trade brought some new life to the sugar plantations in the Northeast, but nothing compared with their apogee in the past centuries. Cotton also flourished in the same region but could never compete with the plantations in the southern United States.[1]

---

1. For economic conditions in the period, see Simonsen 1962 and C. Prado Jr. 1967. On cotton, see Stein 1957.

As the external and more dynamic sectors of the economy shrank, Brazilian life retreated to self-reliance and isolation in the old farms and dormant villages linked only by the slow pace of mule troops, while an unstable political elite worked to consolidate position in Rio de Janeiro. Conflicts between Brazilian elites (mostly regional) and Portuguese elites flared from the beginning, and in 1830 Pedro I abdicated and went to Portugal, where he would be crowned later as Pedro IV. From 1830 to 1840 the country was governed by a succession of regents who ruled in behalf of his son, who was crowned Pedro II in 1840 at the age of fifteen. Pedro II was to remain in power until the beginning of the Republic in 1889. In the regency period a series of regional upheavals threatened to destroy the country's political integration.

The second half of the nineteenth century brought a contrasting period of political consolidation and increasing economic and demographic expansion. For almost fifty years Brazil functioned as a stable parliamentary monarchy, based on limited suffrage and a two-party system in which parties alternated in government under the benevolent supervision of the emperor, who embodied a *poder moderador* (moderating power) in addition to the usual three. The country's provinces, a legacy of old colonial administrative divisions, were governed by envoys from Rio de Janeiro who never remained in their posts long enough to create local links and loyalties, and parliamentary elections were routinely manipulated by the center to ensure their loyalty to the ruling party.

Economic expansion was due mostly to the growth of coffee as an important commodity in the international market. Coffee became an important crop in Minas Gerais and Rio de Janeiro in the early nineteenth century, and boomed due to the availability of cheap land and slave labor. As the land became exhausted, the coffee plantations moved to the South, and at the end of the century São Paulo had replaced the other provinces as the main producing region. This change in geography coincided with a growing scarcity of labor. In 1850 Brazil finally yielded to British pressure and ended the African slave trade. As the slave population dwindled, immigration from Europe and Asia and the substitution of slave work for salaried work emerged as the natural solutions to the crisis.[2]

Demographic and social changes can be explained only partially by these changes in the economy. The occupation of Brazilian territory since the sixteenth century proceeded for a variety of reasons, from military concerns to the presence of Indian populations amenable to enslavement and religious conversion, from the presence of extractive

---

2. See Balán 1973 and Graham 1973 for comparative views on European migration to Brazil.

products to the availability of safe harbors and easy routes to the interior. Brazil was ruled from Salvador, Bahia, until the middle of the eighteenth century, and for most of the colonial period an independent administration also ruled the northern part of the country from the cities of São Luís in Maranhão and Belém at the mouth of the Amazon. Recife had also been the seat of the Dutch colonial adventure in South America, and it long remained the natural outlet and entrepôt for the sugar economy in the Brazilian Northeast. The Portuguese and Spanish empires met and fought for their limits around the Rio de la Plata, and the province of Rio Grande do Sul, with its tradition of military mobilization and insurrections, was in part a product of this conflict. São Paulo was a door to the countryside, a source of Indian slaves and an early seat of Jesuit missions. The discovery of gold in Minas Gerais in the eighteenth century concentrated the country's population in that province, which remained as a peculiar kind of urban/rural society based on slave labor that survived the gold rush a long time. More generally, each administrative or economic cycle left its trace of urban settlements, associated institutions, and population, which led to a complex and urbanized society that coexisted, sometimes with weak or almost no integration, with the plantation economy.[3]

This brief outline should be enough to suggest that science and technology could hardly be expected to emerge in Brazil from the requirements of colonial or postcolonial economic conditions. What we see, instead, are the repeated attempts by the Portuguese and later Brazilian authorities to start some kind of practically minded institutions, followed shortly by their decay or transformation into some kind of unexpected research or general education body. This unexpected and unplanned change should be understood in terms of the modern culture that was beginning to develop in the country's capital in part because of the intellectual Europeanization of some sectors of the Brazilian elite and in part because of a growing number of Europeans—not only from Portugal but also from France, Germany, and other countries—who were attracted by the opportunities of employment or adventure they hoped Brazil could offer.

Colonial Science: The Naturalists

Unlike such nations as France, Holland, and England, which transferred some kind of "colonial science" to the territories and nations they occu-

---

3. See Martins Filho and Martins 1983.

pied, Portugal did not develop its own scientific tradition and therefore could not be expected to take this course.[4] Portuguese colonialism was exploitative and predatory, with no intention of creating in the New World a complex society boasting institutions for producing and transmitting knowledge.[5] Some forms of technology were developed for gold mining and sugar production—the main economic activities during Brazil's four centuries of colonial occupation—and little else.[6]

Brazil was not completely isolated from the rest of the world, however. Throughout the sixteenth century Portugal disputed its possession of the Brazilian territory with other European sea powers, and from 1630 to 1661 Holland controlled the most profitable region of Brazil, the Northeast, from the city of Recife.[7] The Dutch administration brought with it individuals dedicated to the study of Brazil's geography, zoology, and botany and left behind an important collection of drawings that is only now being rediscovered. Scientific activities undertaken in Brazil until independence were to focus on descriptions of nature within the New World: its fauna, flora, minerals, and inhabitants. It was descriptive science, undertaken largely by foreign travelers, who added to the observations on natural history then being accumulated in Europe.

The Portuguese crown's interest in Brazil's raw materials led to some efforts to collect information about new products of possible commercial value. Brazilian scientific and educational institutions up to the second half of the nineteenth century cannot be compared with those in Spanish America. Education, under the direction of the Jesuits, never exceeded the equivalent of today's secondary school. Leery of the idea of Brazilian institutes that might rival those in Portugal, the crown hindered the Jesuits from establishing their proposed university and prevented the creation of any kind of press that might have contributed to the dissemination of new ideas.

With Pombal's rise in Portugal, the colonial scene underwent substantial changes. In 1783 the Portuguese government entrusted Alexandre Rodrigues Ferreira, the first Brazilian naturalist to have studied at Coimbra, with the responsibility of exploring the colony's flora and fauna. An important contribution to Brazilian botany and zoology, the results of Ferreira's studies were all lost for Portugal during the invasion of Portugal by Napoleon's troops, when the holdings of the Real Museu were taken to Paris by Geoffroy de Saint-Hilaire.

4. McLeod 1975; Moyal 1976.
5. Godinho, 1961–70; Lang 1979; Maxwell 1972.
6. A. de B. Castro 1971.
7. Melo 1976; Boxer 1965 and 1973.

In 1772, under the viceroyalty of the Marquis of Lavradio, a Sociedade Científica was founded in Rio de Janeiro for the purpose of disseminating scientific knowledge. The society sponsored public conferences and dealt with a range of subjects: botany, zoology, chemistry, physics, and mineralogy. The Marquis de Lavradio also created a small botanical garden for plant experimentation. In 1779 the society's name was changed to Sociedade Literária do Rio de Janeiro. Its work continued until 1794, when it was closed—probably for minor political reasons.[8] In 1797 the first official research institution was finally installed in Brazil, when the Portuguese king ordered the captain general of Pará to set up a botanical garden for the acclimatization of plants in the city of Belém.[9]

Imperial Science: The Nineteenth Century

It was only in the nineteenth century, following the transfer of the Portuguese court to Brazil, that some technical institutes and more systematic research activities began to appear. For geologist Othon Leonardos, Brazilian science had its real start with the brothers Martin Francisco and José Bonifácio de Andrada e Silva, who traveled to rural São Paulo in 1819 to study its geology and mineralogy and then to apply this knowledge to mining activities.[10] Several institutions had been created already in 1808: the Academia de Guardas-Marinha in Rio de Janeiro, later to become the Naval Academy; Bahia's Colégio Médico-Cirúrgico and Rio de Janeiro's Escola Médico-Cirúrgica, which developed into the country's first two schools of medicine; the Biblioteca Nacional, Brazil's national library; Rio de Janeiro's Jardim Botânico, originally known as the Real Horto; and the Escola Central, a military academy that was to become the country's first engineering school.

The pragmatic intent of these early institutions and how they drifted away from it are clear in the examples of the Jardim Botânico and the Museu Imperial.[11] The Jardim Botânico had its origin in the establishment of a gunpowder factory near the city's Rodrigo de Freitas lake. Parallel to the founding of this factory on 13 June 1808, the prince

---

8. Azevedo 1885; Alexander Marchant 1961; Alden 1968.
9. Anyda Marchant 1961.
10. Leonardos 1955:271.
11. Another institution from those years was the Laboratório Químico-Prático (Laboratory of Practical Chemistry), established by João VI in 1812, which H. Rheinboldt considered the site of Brazil's first chemical-industrial operations. See Rheinboldt 1955:23–25.

regent[12] decreed that a lot be prepared near the plant inspector's lodgings where an acclimatization house for East Indian spices would be built. Besides growing East Indian spices, the garden was to be used to acclimatize and cultivate tea for the European market. In 1814 a group of Chinese colonists settled in the region and demonstrated how to prepare the product. Although the crop did reasonably well, the original plan of supplying Europe was never implemented. The Jardim Botânico served simultaneously as the main site for developing and acclimatizing such plants as the nutmeg, avocado, clove, cinnamon trees, sugarcane, and other plants. Its example spread, and botanical gardens were set up one by one in Bahia, Minas Gerais, Pernambuco, São Paulo, and other places using plant seeds and seedlings originally sent to Rio from abroad. Later, King João expanded the Real Horto, opening it to the public and renaming it the Real Jardim Botânico. Under the administration of Friar Leandro do Sacramento, the first professor of botany at the Escola Médico-Cirúrgica and the first director of the Jardim Botânico after independence, the garden's initial role was significantly expanded. From a mere lot for the introduction and acclimatization of plants, it grew to an institute for serious experimentation and study. Besides initiating the cultivation of some plants, including tea, Friar Leandro's administration exchanged species with Cambridge's botanical gardens and distributed seeds and plants to gardens in Pará, Pernambuco, and Bahia. As the economic irrelevance of these products became obvious, the garden turned into a place for traditional studies in botanical taxonomy—and mostly a pleasant park and strolling ground for the population of Rio de Janeiro.

The Museu Real (later to be named Museu Imperial and later still Museu Nacional) started with a collection of mineral samples from the German mineralogist Abraham Werner which was then being used in practical classes at the Academia Militar; art objects in wood, marble, silver, ivory, and coral; a collection of oil paintings donated by King João VI; native handicrafts and natural products dispersed among various Rio de Janeiro establishments; and stuffed animals from an old collection started at the time of the colony and known as the Casa dos Pássaros (Bird's House).[13] In addition, many private donations were made. The original

---

12. The Portuguese court that fled to Brazil in 1808 was headed by Prince Regent João, who was ruling in behalf of the deranged mother queen. He was later crowned as João VI.

13. Portugal's utilitarian interest in Brazil was evident in the decree that created the Museu Real on 6 June 1808: "In the interest of propagating the knowledge and study of the natural sciences in the kingdom of Brazil, which encompasses thousands of objects worthy of observation and examination and which may be useful to commerce, industry, and the arts (all of which I would like to bless with great fountains of wealth): I hereby

administration included a director, a janitor, a zoological assistant, a clerk, and a bookkeeper. A total of 2,880 *mil réis*, the equivalent of approximately 580 British pounds, was first budgeted for the purchase of material.

The Museu Nacional developed slowly. It was some time before the organizing of public exhibits of its collections became an important or even permissible activity. Until 1821 only two rooms on the ground floor of the Campo de Santana building were opened to the public; there a display of models of industrial machinery had been set up on the initiative of another institution, the Sociedade Auxiliadora da Indústria Nacional. That year the museum's scientific collections were opened to visitation. After independence in 1822, the museum entered a new and fruitful period. The ministers of the Empire supported the museum by requesting that the foreign naturalists who came to Brazil donate collected material. The museum slowly expanded its collections with donations from Langsdorff, Natterer, Sellow, and others. It was then that a laboratory for physics and chemistry was set up, and the museum commenced systematic exchanges of collections, or samples of collections, with foreign institutions.

As the century progressed, the Museu Nacional developed into a scientific center where European naturalists would gather on their arrival in Brazil. Ludwig Riedel, who came to join the scientific expedition of G. I. Langsdorff[14] in 1820, served as head of the botany section of the museum for some time, and Friedrich Sellow, who had also come to Brazil at Langsdorff's suggestion, spent time traveling on commission to the museum. Fritz Müller, whose *Für Darwin* is considered a contribution to evolutionist theory, was a traveling naturalist for the museum for many years. Other memorable names include Hermann von Ihering and Émil Göldi. Serving as traveling naturalist for the museum following his ar-

---

determine that a royal museum be established in this court, to which place the instruments, machinery, and offices presently scattered in other locations shall be transferred as soon as possible, all under the responsibility of those whom I choose to appoint in the future. And being it of my knowledge that the dwelling places now occupied in the Campo de Santana by their owner João Rodrigues Pereira de Almeida are of adequate proportions and rooms for this purpose, and that the aforementioned owner has voluntarily agreed to sell this property for the sum of thirty-two *contos* in service to me, I have decided to accept this offer, proceeding with the transfer of deed through the Board of Finance so as to incorporate this property into the crown's possessions." (Quoted in Lacerda 1905:3–4.)

14. Langsdorff, of German descent, was designated as the Russian consul in Brazil. Between 1820 and 1827 he organized two expeditions through most of Brazil's interior, collecting a large quantity of botanical, zoological, and ethnographic materials that were sent to Saint Petersburg and are only recently being opened for study. See Chur, Komissarov & Licenko 1981.

rival from Germany, von Ihering became the founder and first director of São Paulo's Museu Paulista in 1894. Göldi joined the Museu Imperial as an aide in the zoological section and was later invited to organize the Museu do Pará, which now bears his name.

In 1876 the Comissão Geológica do Império (Imperial Geological Commission) provided the museum with a major geological collection organized by Charles F. Hartt. Gold, silver, iron, copper, lead, tin, and precious gems are also part of these collections, which contain samples from Mexico, the United States, Russia, Austria, and other nations. As of 1850 the museum's well-equipped mineralogy laboratory broadened its analyses and experiments to include fuel samples. Foreign explorers were focusing on discovering deposits of coal, petroleum, copper, and other minerals,[15] but the museum's role was to be reduced by the transfer of responsibility for the main geological and mineral activities to Brazil's geological services.

The museum reached its golden age in 1876 under Ladislau Neto:

> There was enthusiasm in the air, a desire to build the museum's reputation and gain esteem in the eyes of the public and of the nation's government. Work was undertaken earnestly in laboratories and offices; collections were reexamined, and old or damaged specimens were substituted with more recent ones; cupboards were filled; scattered bones were brought together to form skeletons; hides were finally put to use; care was taken concerning the aesthetic appeal of collections on exhibit; labels replaced with new ones; and old generic denominations were substituted by their modern adaptations. . . . Conferences held in the museum's salon at night attracted a distinguished and select group, which most often included the lively presence of Emperor Pedro II. On scheduled dates, teachers, deputies, senators, top civil servants, and ladies of high society all gathered there to hear a fascinating and useful lesson on any of various branches of natural science illustrated with drawings and prints, murals, and samples of the objects mentioned in the lecture. The topics of zoology, botany, and biology were all approached synthetically, and the lecturer would provide the audience with conclusions and a summary of facts that were easy to retain and assimilate. Papers from these conferences were published in the newspapers and in certain literary and scientific journals. Open to the public three days a week, the museum drew thousands of visitors a month eager to view the objects on exhibit. . . . Everywhere, the museum was spoken of

---

15. Lacerda 1905:26–27.

fondly and praised. Travelers who stopped to see the capital of Brazil were anxious to see its collections.[16]

In 1880 the museum opened a laboratory of experimental physiology—the first in the country—where João Batista de Lacerda and Louis Couty were to carry out their work. Couty had come to Brazil from France to teach industrial biology at the Escola Politécnica, but he turned to the museum as an adequate place for practical experimentation. The first studies involved animal poisons; toxic and nutritive plants; climate physiology; sugarcane alcohol, coffee, and mate; human and animal diseases; and cerebral physiology, using monkey subjects. All students of the development of biological science in Brazil stress the importance of Couty and Lacerda's laboratory as long as it lasted.[17]

By the beginning of the twentieth century the geology and mineralogy sections of the museum had accumulated a significant collection that included samples of nearly all minerals then exploited in Brazil. But this was already a period of institutional decadence. As the Republic brought new urgencies and priorities, it could not find a place for the old museum, which became mostly a lifeless collection of scientific curiosities for the eventual visitor.

The apogee of imperial science was marked by the active presence of Pedro II himself in all matters dealing with science, technology, and education. Playing the role of a Brazilian Maecenas, the emperor's attraction to the sciences led him to seek the company of scientists both within Brazil and abroad and to participate in all of Brazil's more significant cultural and scientific events.

The emperor's direct involvement with the sciences met with some resistance. Fernando de Azevedo interpreted this as a resistance to modernization, the "poorly disguised hostility felt within an intellectual and political environment that was dominated by individuals prone to rhetoric and educated in abstractions—an environment that led national thinking to absorb itself in literature, legal concerns and questions, and political debate."[18] Besides considerations of this nature, the emperor's interest in scientific matters placed such activities at the mercy of imperial whims. Those who believed they were not getting a fair deal were perhaps in the best position to realize the dangers inherent in this situation. Such was the case of Joaquim Murtinho, a homeopathic physician who in his defense of homeopathic medicine spoke out sharply against Pedro II:

16. Lacerda 1905:44–45.
17. L. de C. Faria 1951.
18. F. de Azevedo 1963:395.

His royal highness suffers from what we might call scientific mania. His royal highness, with one single thesis in hand, wants this thesis to encompass medicine, mathematics, natural sciences, civil and mining engineering, philosophy, history, language, ... and whatever else; his royal highness studies this thesis most seriously and, whenever a defendant mentions a certain point in their work, his royal highness leafs through the thesis as if he understood the subject matter and were trying to formulate an opinion.... Be it a scientific or industrial experiment, an attempt to guide a balloon, an experiment on electrical lighting—there is his majesty quoting the books he has read on the subject and voicing his opinion on the results of the experiment.

The emperor not only had opinions but also made decisions:

When there is a selective examination for candidate teachers in our schools, off go the candidates' exams to be read by his royal highness. And fortunate are those whose exams please his majesty. When it comes to hiring a foreign professor for a position at one of our universities, it is not the faculty that advises the government about the candidate with the best curriculum—it is his royal highness himself, or one of his scientific aides, who selects the appointee. Physiologists are sent to teach agriculture, and mining engineers are sent to teach the arts and manufacturing, ignoring professional callings, displacing individuals from their chosen areas, and transforming professors who are distinguished in their fields into mediocre teachers who must teach on subjects they are not familiar with simply because his royal highness chose to put them there. In all his acts, his royal highness seems to say: Science, is me.[19]

Higher Education

The authorities were involved not only in science but also in education. The distribution of new educational institutions in the Brazilian territory early in the nineteenth century tells us something about their purposes. The transfer of the Portuguese crown to Brazil was the lowest point in the history of the Portuguese Empire since the glorious years of the discover-

19. Quoted in Lobo 1964: vol. 3.

ies, and Brazil's military weakness explains the priority given the establishment of military schools in the capital, Rio de Janeiro. The second priority was medicine and surgery, both for military reasons and supposedly for the protection of health. After the capital, Bahia was Brazil's largest and most important city, and it was fitting that Bahia should be the seat of the second medical school. A law education was probably still seen as the best mobility channel for the children of the local gentry in decadent Recife and stagnant São Paulo, and they got the law schools they longed for.

If these were the motivations of those surrounding the exiled Portuguese king, the pattern did not remain unchanged or unquestioned in the years to follow. The military academies developed into engineering schools that did not excel as technical centers but that did provide fertile ground for the scientific values of positivism; and the medical profession, stimulated by its newly discovered efficacy against tropical diseases at the turn of the century, also developed its own ambitions. The law schools, established in 1827 in São Paulo and Recife, moved away from the dominance of canonical law and the traditional Portuguese codes and received an influx of different strains of European liberal thinking.[20]

The Empire's educational system was characterized chiefly by administrative centralization. According to the Royal Charter of Law of 4 December 1810, the Academia Real Militar was to be "headed by a military junta, composed of a president and four or more deputies, three of which were to be those whom, as more capable in scientific and military studies, I decide to select and appoint to said position."[21] All the king's appointees in an 1811 decree were members of the Royal Engineering Corps.[22] Even after the introduction of a selective examination scheme for the Academy in 1833, politics continued to play a large part in the hiring of faculty. A 1837 report on the state of professional teaching in Brazil called attention to "the poor choice of some faculty members, appointed under scandalous favoritism. Instead of seeking out the most worthwhile candidates, with a few honorable exceptions an effort has

---

20. On the law schools in the nineteenth century, see Venâncio Filho 1977 and Adorno 1988.
21. Quoted by F. M. de O. Castro 1955:50.
22. Antônio José do Amaral, first lieutenant, native of Rio de Janeiro, and first-year lecturer; Francisco Cordeiro da Silva e Alvim, master sergeant, native of Portugal, and second-year lecturer, later to become the Viscount of Jerumirim; José Saturnino da Costa Pereira, first lieutenant, native of Colônia do Sacramento, located in Brazil's southernmost state of Rio Grande do Sul, and third-year lecturer; Manuel Ferreira de Araújo Guimarães, captain, native of Bahia's capital city of São Salvador, and fourth-year lecturer; and José Vitorino dos Santos e Souza, second lieutenant, birthplace unknown, and lecturer of descriptive geometry (Morais 1955:118; F. M. de O. Castro 1955:52).

been made to select only protégés. . . . Favoritism in selective examinations has been such that it is repulsive even to speak of it: sons follow fathers, brothers-in-law follow brothers-in-law, cousins follow cousins, nephews follow uncles. . . ."[23] Without significant demand for qualified professionals and for a professional community that could and would impose some standards of quality, it was inevitable that centralization should have such negative effects. The result was that the Academy operated inadequately and presented many deficiencies: a lack of diligence by students and faculty, constant cheating on exams, and carelessness in the preparation of textbooks.

The very books the professors had to use in preparing textbooks were also prescribed by law. Centralization was also apparent in the direct subordination of the schools to the imperial cabinet. All institutions had to follow a mandatory seven-month school year, and "preparatory exams" before Provincial Committees of Public Teaching were established for candidates for higher educational institutes later in the century. These exams were based on subjects taught in secondary schools— almost solely humanistic—and deprived schools of the right to choose students according to their own criteria.[24]

Brazil was to witness a profound turnabout in its higher education system through the Leôncio de Carvalho reform, carried out in imperial times under the administration of Prime Minister Viscount of Sinimbu. This transformation was prompted by a vague awareness of the German university system mixed with positivist thinking and adapted to the political climate of decentralization that prevailed since the 1870 republican manifesto. Attendance was made optional and freedom of teaching was adopted with the introduction of a Brazilian version of German *privatdozent* and the elimination of government control over what was taught at the schools. According to contemporaries, the effects were disastrous. Whatever little quality control existed in the previous centralized regime immediately ceased to exist. A system of qualifying exams provided by the state at the end of the courses was introduced to compensate for the lack of controls, but its reliability depended too much on the capacity of each particular teacher.[25] The main result of this legislation, which was in effect until 1895, was the spread of institutions of higher education throughout Brazil, starting in São Paulo. The Leôncio de Carvalho reform left the impression that

23. Clóvis Beviláqua, quoted in Almeida Jr.: 1956:21, 22.
24. J. M. Carvalho 1978.
25. Almeida Jr. 1956; Venâncio Filho 1977; Barros 1959.

Brazil was not ready for academic freedom and pluralism, which reinforced the predominantly authoritarian and centralizing tendencies that were to prevail to the present.

With these limitations and lack of autonomy, the higher education institutions became the main centers from which the early Brazilian traditions of scientific work in the physical and biological sciences would be established.

Engineering and Mines

The first technical institution in Brazil was the Academia Real de Marinha (Royal Navy Academy), created by King João VI within Rio de Janeiro's São Bento Monastery. Two years later that same city gained its Academia Real Militar (Royal Military Academy), which was responsible for training officials in artillery and in geographical and topographical engineering. Enacted on 4 December 1810, the military academy's Charter of Law stated that instruction was to consist of "a complete course in the mathematical sciences, in the sciences of observation—that is, physics, chemistry, mineralogy, metallurgy, and natural history, including the vegetable and animal kingdoms—as well as all military sciences, including both tactics as well as construction of forts and artillery."[26] In 1832 the two academies were joined to form the Academia Militar e de Marinha, but that union did not last more than a year.

The Academia Militar's complete course was seven years long and divided between mathematics (four years) and military teaching (three years). The teaching of mathematics followed these lines:

> The first-year lecturer taught arithmetic, algebra (up through third- and fourth-degree equations), geometry, linear trigonometry, and some basics of spherical trigonometry; the second-year lecturer taught advanced algebra, analytical geometry, and differential and integral calculus; the third-year lecturer taught mechanics (static and dynamic), hydrostatics, and hydrodynamics; and in the fourth year there was a lecturer on spherical trigonometry, optics, astronomy, and geodesics.[27]

26. Quoted in F. M. de O. Castro 1955:56. See also, on Rio de Janeiro's engineering tradition, Barata 1973.
27. F. M. de O. Castro 1955:51.

This program was structured so that its concepts would be rigidly applied and leave no room for doubt or experimentation.[28]

All further reforms within the Academia Real Militar either dealt with questions of a disciplinary nature or sought to improve the purely professional aspects of its courses. Later, civilian and military engineering began to move apart from each other. Beginning in 1833, civilians were allowed to attend courses with members of the military. In 1839 the academy became the Escola Militar and soon earned a reputation for maintaining extremely strict discipline and therefore no longer attracted the interest of civilians. But in 1842 a seven-year engineering course for civilians was introduced, and it became possible to obtain a doctoral degree through presentation of a thesis. An Escola de Aplicação (Practical School) was instituted in 1855 to make independent the teaching of military subjects, and in 1858 the Escola Militar was renamed the Escola Central as a predominantly civilian institution. Members of the military continued to attend, taking classes common to both courses. Physics was taught as a separate subject as of 1858.[29]

In 1874, under the cabinet of Visconde do Rio Branco, Brazil's higher education system was thoroughly reformed, and the civil and military engineering courses were effectively separated. This resulted in the creation of Rio de Janeiro's Escola Politécnica along the French model. At this time the imperial regime was in full bloom, population was expanding, coffee was providing new revenues for the landed gentry and new taxes for the central government, contacts with Europe were intensified, and the old educational institutions became too narrow for the sons of an

---

28. For instance, the second chapter of the mathematics charter reads: "The fourth-year lecturer shall explain the full extent of Lagrange's spherical trigonometry, as well as the principles of optics, catoptric, and dioptrics; the basics of all types of spectacles (refraction and reflection) shall be presented, followed by an explanation of the system of the world, for which the works of Laplace shall be most helpful—not venturing into his noble theories, as time will not permit as much, but revealing the main results so elegantly demonstrated by Laplace, explaining all methods used in determining latitudes and longitudes at sea and on land; and regularly commenting and demonstrating how this may be applied to geodesic measures, once again to the fullest extent possible. The lecturer shall reveal also the basics of geographical maps, of various projections, and application to geographical maps and topographies, as well as explain the principles behind reduced maritime maps and the new method used to draw up the map of France; also providing a general idea of global geography and its divisions. The works of Laplace, of Lacaille, and the Introduction of Lacroix and the geography of Pinkerton will serve as a basis for the textbook to be compiled and within which an effort must be made to cover the full extent of these topics" (quoted in Morais 1955:117).

29. F. M. de O. Castro 1955; Morais 1955; Ribeiro 1955; Almeida Jr. 1956.

expanded elite. In the new Politécnica the old course of mathematics at the military school was divided into one course of physical and mathematical sciences and another of physical and natural sciences. Within physical and mathematical sciences classes were offered in "celestial mechanics and mathematical physics" and "supplementary mathematics." Also, in an important and unprecedented change, degrees could be attained as bachelor or doctor of physical and mathematical sciences or of physical and natural sciences outside the professional engineering courses. Science was thus introduced in the school, leading to the high expectations Rio Branco held about the Escola Politécnica; in an 1876 report he described the new curriculum as embodying "the culmination of advances attained by natural and physical-mathematical sciences."[30]

This was too optimistic. Neither the spirit nor the structure of these courses remained during the republican period, after 1889. The first attempt to do away with these courses came in 1890 under the provisional government, soon after the fall of the Empire in 1889. The faculty at the Escola Politécnica spoke out against the proposed reform, so the head of the provisional government decided against carrying it out. The Politécnica's scientific courses survived until 1896, when they were finally abolished by the school's own faculty.[31]

Ouro Preto's Escola de Minas stands as a remarkable exception among the professional schools that came into existence following Rio Branco's educational reform. Created in 1875 by the personal initiative of the emperor, the Escola came to life under Claude Henri Gorceix, its organizer and first head.[32] During a visit to Europe at the beginning of the 1870s, Brazilian Emperor Pedro II invited Auguste Daubrée, director of the Paris École de Mines, to organize and head a Brazilian counterpart; instead, Daubrée suggested Henri Gorceix for the post. In July 1875, one year after his arrival in Brazil, Gorceix presented the government with a report suggesting the future location and the bylaws for the Escola de Minas. It was to be built in the colonial town of Ouro Preto, seat of the Minas Gerais province, close to the country's richest mineral deposits. He demanded that it be a two-year course, with classes running for a ten-month period from August through June, the remaining two months being dedicated to excursions and practical work. The course would be full-time for both professors and students, with the former receiving good salaries and the poorer students receiving free classes and scholar-

30. Quoted in F. M. de O. Castro 1955:61.
31. F. M. de O. Castro 1955:62.
32. For a full account, see J. M. Carvalho 1978.

ships. A maximum of ten students per class would be allowed, and those who performed best would be awarded free trips to the United States and Europe for further improvement. An entrance exam would determine admittance independent of official Committees for Public Teaching, which controlled the access to other higher education institutions in the country, and exams would be frequent during the course itself. Last, the government would single out and make use of those who performed best during training trips abroad. Enacted on 6 November 1875, after a few changes regarding expenditures, this initial project met with Daubrée's full approval.

The final project of the Escola de Minas was inspired not by the famous Paris school but by Saint-Etienne's. The nature of teaching at the former was more general and attracted the best graduates of the École Polytechnique to its three-year course; Saint-Etienne's two-year course was more practical and operational, although it did seek to provide better education than what would be required for simple technicians or master craftsmen. The Ouro Preto school was to be an *école de mineurs*, not an *école de mines* in the Paris tradition.

From its beginnings the history of the Escola de Minas was marked by a continual struggle against the imperial cabinet's centralizing tendencies and by constant clashes with Rio de Janeiro's Politécnica over its status, autonomy, and goals. An 1880 decree determined that graduates of the Escola de Minas should receive equal treatment when competing for teaching positions at similar schools. In 1885 the Ouro Preto course was granted the same status as the engineering course offered by the Politécnica. Despite this legal guarantee, however, Politécnica examining boards always managed to reject Ouro Preto graduates applying for professorships. On several occasions the personal intervention of the emperor was necessary. The lack of a specialized market for the school's graduates made it necessary to include civil engineering in the course, as recommended in 1884 by the president of the province of Minas Gerais, who offered his support to the school just as the central government's funds were becoming scarce. According to J. M. Carvalho, "the intervention of the province, which demanded changes in the initial project, most likely saved the school from extinction. But this intervention arose not from an interest in preserving an upper-level Escola de Minas but from a desire to preserve an upper-level school of any type in the Province of Minas."[33]

---

33. J. M. Carvalho 1978:59.

Medicine and Surgery

The 1808 initiation of two medical-surgical courses, one in Salvador and the other in Rio de Janeiro, marked the official inauguration of medical teaching in Brazil. Before that time, medical assistance in the colony was provided either by herbal *curandeiros*—inheritors of empirical, indigenous, or African knowledge—or by practitioners working under Portugal's Proto-Medicato. The Proto-Medicato was a permanent board that supervised all practices related to the art of healing; it also passed judgment on and submitted for official approval formal requests for authorization to practice these activities. To be so certified, the candidate needed to present a declaration attesting to a certain period of apprenticeship under another professional and to pass a brief exam before this medical board.[34]

In 1808 the Portuguese crown created in Brazil the posts of Físico-Mór do Reino (physician general of the kingdom) and Cirurgião-Mór do Exército (surgeon general of the army), which became the highest sanitation authorities within Portugal's administrative organization. Together these two posts formed a kind of board of public health. The surgeon general and his delegates were to oversee everything related to teaching and to the practice of surgery, bleeding, birth, tooth-pulling, the application of leeches, and bone-setting. Besides their responsibility over military hospitals, doctors, and medical services, the physician general and his delegates were to oversee the teaching and practice of medicine; questions arising between doctors and patients; the practice of pharmacies, apothecaries, druggists, *curandeiros*, and surgeons dealing with internal illness. They would also be responsible for preventing epidemics and for supervising urban sanitation. The hierarchy between medicine, a liberal profession, and surgery, a practical skill, was obvious.

As the kingdom's new surgeon general, José Correia Picanço, a native of the state of Pernambuco and a graduate of Coimbra, suggested the creation of the Escola de Anatomia e Cirurgia (School of Anatomy and Surgery) in Bahia. The school was to operate in Bahia's Hospital Real "in benefit of the preservation and health of the citizens, with the purpose of training able and expert professors who, through the union of medical science and practical surgical knowledge, could be of service to residents of Brazil."[35] Rio de Janeiro's course was created a short time later, due to

---

34. The following account is based on Magalhães 1932; Campos 1941; Lobo 1964, 1: chap. 2; Santos Filho 1947 and 1977; and Lacaz 1977.
35. Quoted by Lobo 1964, 1:13.

"a dire need for the court's military and marine hospitals to train their surgeons in the principles of medicine, as well as to treat the ailing aboard ships and the people who must reside in distant villages of the vast continent that is Brazil."[36] Four classes were offered during the four years of study: anatomy and physiology; surgical and private therapy; surgical and obstetrical medicine; and medicine, chemistry, medical topics, and pharmacy. Upon completion of studies, the student received a certificate and could then request that the surgeon general convene an examining board to judge the candidate's qualifications. After the graduate received a diploma, approval also would have to come from the Universidade de Coimbra. An 1811 reform, based on the Coimbra model, would require that, to be accepted, a candidate should have knowledge of Latin; rational and moral philosophy; geometry; and some elements of algebra, physics, and chemistry. This proposed curriculum was much broader than what is common today. The pharmacy course was to take three years; surgery and medicine, five years. It was never implemented.

In 1813 Rio de Janeiro's medical school was reorganized along much less ambitious lines and renamed the Academia Médico-Cirúrgica. Now focused on surgery, the course program excluded pharmacy and medicine. To gain admittance, a candidate needed only to read and write Portuguese correctly and agree to learn French and English during the course itself. Students already knowledgeable in Latin or geometry were allowed to skip the first year altogether. After attending the academy for five years and having been approved in all final exams, a student got a Letter of Approval in Surgery. Those who continued their studies for another two years received a Letter of Graduate in Surgery, which guaranteed its holder various privileges: preferential placement for public job openings; permission to treat all diseases in places where there were no medical doctors; automatic membership in the Colégio Cirúrgico; and participation in the Academy of Medicine in Rio de Janeiro and in all similar institutions still to be established in Brazil. The degree of Doctor of Medicine would be bestowed on any graduating surgeon who presented a dissertation in Latin and successfully completed the exams indicated by the surgeon general. In 1815 the Bahia School was reorganized along the same lines. The structure of the schools of medicine was not changed because of Brazil's 1822 formal independence. Not until 1826 was the need for degree confirmation by the Universidade de Coimbra eliminated.

The 1829 founding of the Sociedade de Medicina (Medical Society)

36. Lobo 1964, 1:13.

was a sign of its growing prestige and professionalization. The first nucleus of this society was a group of five distinguished physicians—two Brazilians and three foreign—plus two graduate surgeons. Organized according to the French Academy, its first task was to study the projects for the reform of medical teaching then under discussion by the Congress. Approved by the Congress after a few changes, and enacted on 3 October 1832, the project raised the status of the schools of Bahia and Rio de Janeiro to schools or colleges of medicine, authorizing them to grant diplomas of Doctor of Medicine, Pharmacy, or Midwifery (the diploma for "bleeder" was eliminated).

This reorganization of medical teaching was supposed to mark the passage from symptomatological and practical medicine to scientific medicine. The reformers criticized the old curriculum because it did not

> offer one single course within the so-called auxiliary sciences dealing with the study of nature or of bodies and of the general and specific properties of each. . . . Physics, chemistry, and botany—these sciences are indispensable to the study of medicine; they provide us with innumerable documents that can be used either to explain phenomena of the organism or to examine the composition and action of bodies or to search for medicinal and mechanical means of protecting health and curing disease.[37]

Besides including the three auxiliary sciences, this new course plan highlighted and expanded the teaching of hygiene, a field that was to be especially emphasized in Rio de Janeiro. Nonetheless, clinical practice remained the strong point at both schools.

An exception to this clinical tradition was the so-called Escola Tropicalista Bahiana, which was not a school in the formal sense but a movement that started around 1850 and developed outside Bahia's Escola de Medicina. Except for Otto Wücherer and John Ligertwood Paterson, who were educated abroad, all the members of the Tropicalista movement studied at the Escola de Medicina.

The Escola Tropicalista Bahiana made significant contributions. Wücherer and Paterson identified the yellow fever epidemic in 1849 and cholera morbus in 1853. In 1863 Wücherer published an essay on Brazilian fauna examining and describing new species of snakes and establishing morphological rules for the identification of poisonous varieties. He was also responsible for the correct identification and description of several diseases, including the hookworm infection, while Silva Lima

---

37. Quoted by Lobo 1964, 1:50.

described beriberi more precisely than ever before. The work of the Escola Tropicalista Bahiana is recorded in the *Gazeta Médica da Bahia*, which began publication in 1866.[38] Considered a very good journal for its time, it appeared regularly until 1908 and served as a vehicle for the dissemination of work by other members of the Tropicalista movement. Available literature does not shed much light on the possible nature of the relationship between the movement and the Escola de Medicina—whether it was one of collaboration or of rivalry—but it is difficult to imagine how one could ignore the other in nineteenth-century Salvador. It is likely that the model of teaching and research that was to permeate Brazilian science for another hundred years—that is, that the two should be carried out in separate locations—was already taking root.

Imperial Science in Perspective

Scientific activity until the beginning of the Brazilian Republic was extremely precarious. On the one hand, it had to deal with unstable initiatives undertaken at the emperor's fancy; on the other hand, it had to deal with the limitations of bureaucratic professional schools that had no autonomy and had completely utilitarian goals.

This precarious situation can be better understood if we remember that Brazil did not have significant social sectors that judged scientific activities worthwhile and important enough to warrant interest and investment. To gain a better perspective, one can contrast it to what was happening at more or less the same time in two non-Western countries of significant size, Japan and India.

Japan had been endeavoring systematically to absorb Western technology and science ever since the Meiji Restoration in 1868. By the year 1900, Tokyo's Imperial University was already offering advanced classes in physics, technology, and the biological sciences, all taught in Western languages. In addition, the Japanese students sent to study at the most important scientific centers in Europe and the United States were later to do the scientific teaching back in their homeland. In taking on this challenge, the Japanese government could rely on the support of a well-defined social group, the Samurai. When the period of feudal decentralization ended in Japan, this traditional warrior class abandoned its former activities and began supplying the individuals needed to accomplish Japan's scientific and technological revolution.[39]

38. A facsimile reproduction in two volumes was published by Falcão (ed.) 1974.
39. Koizumi 1975; Hashimoto 1963.

Although different in many aspects, India by the nineteenth century was also highly involved in Western culture. The English brought their teaching methods to the colony and encouraged the local elite to send their children to British universities. Indian society went through a process of Westernization that eventually led to the adoption of English as the colony's official language. India's cultural elite, the Brahman caste, moved into the new schools and universities hoping to maintain their social and cultural leadership within the limits permitted by their colonizers. Students of Indian history tend to stress how sterile this process of establishing a dynamic scientific and technological attitude that India could call its own was.[40]

This look at nineteenth-century Japan and India underscores the meagerness of the educational and scientific projects being undertaken by the Brazilian Empire. Science was first perceived as applied knowledge, and as such it proved unpractical or uneconomical; later, it was perceived as culture and therefore mostly irrelevant. The gradual expansion of higher education throughout the nineteenth century was in part a quest for new and useful knowledge with a growing scientific content. It was also part of a movement from a limited but growing urban elite to open space and gather recognition in society through the strength of their special asset: the new knowledge they captured in Europe and were carrying to Brazil.[41]

The way the old Rio military school changed its names and goals during the nineteenth century is a good indication of its perceived roles. The military profession never enjoyed too much prestige in Brazil except at the southern frontier, and the civilian dimensions of the school always prevailed. In 1858 the Military Academy changed its name to Escola Central, and finally in 1874 it adopted the French denomination of Escola Politécnica. That civilian engineering was dominant did not mean the school was particularly competent in development of mechanical or construction skills or in the stimulation of competence in the physical and natural sciences. Contemporary visitors were unanimous in criticizing the way teaching was conducted: with outdated textbooks, without practical or experimental classes, and almost totally without independent research work. This was probably just as well for the limited technological needs of Brazilian society at the time. The Escola de Minas de Ouro Preto did not fare much better in the long run, in spite of far stricter standards initially; Minas Gerais' soil was rich, but there would never be an economic basis for a mining industry that needed the skills the Escola de Minas was supposed to develop. Only in São Paulo, where the local Escola Politécnica was created in 1894 to follow closely the

40. Morehause 1971; Rahman 1970.
41. The following is based on Schwartzman 1991.

expansion of the railroad system into the coffee country, was a more technical and specialized education actually achieved.

What gave meaning to the Escola Politécnica in Rio de Janeiro (as well as to the Escola de Minas and to some extent the Politécnica in São Paulo) was mostly their role in the creation of a new breed of elite intellectuals who could challenge the established wisdom of priests and lawyers in the name of modern science. The notion that society could be planned and ruled by engineers, which was well within the French tradition, would have a large impact in Brazil. While in the British tradition engineering has always been a minor and ungentlemanly occupation, the École Polytechnique was since its inception the place where the French elite was to be educated. There, military education came with the training of the mind in mathematics and physics, and it was believed that this combination would prepare the best of Cartesian minds to build bridges, run armies, and rule the economy. Positivist doctrine assured Brazilian engineers that they had the right and competence to rule society and make that society better and more civilized under their rule. They campaigned against the monarchy, for universal education, and for better salaries for the working class; they opposed the church and all forms of corporatist organization (universities, with their ambitions of self-regulation, were seen as one form), opposed mandatory smallpox inoculation, and above all organized themselves into secret societies and conspired to gain power. They were so successful that their slogan, Ordem e Progresso (Order and Progress), is today still on the Brazilian flag.

A similar picture can be found in medicine. The notion that the medical sciences should move from their healing role to a more social, preventive one became well established in Brazilian medical circles in the nineteenth century.[42] Previously, doctors or healers dealt mostly with individuals who sought their help and who could afford to pay for their services. Global epidemics—the plagues, leprosy, the pox, venereal diseases—were to be handled by government and religious authorities—isolating the carriers, comforting the dying, and exhorting the healthy not to live in promiscuity. Early in the century, probably for the first time in Brazil, doctors were asked to explain the causes of the illnesses of Rio de Janeiro as a city and to suggest cures. They found problems with the air, the architecture, the supply of foods to the population, and the social morality. Their recommendations were mostly urbanistic, legal, and moral instead of strictly medical, and they required the approval of higher authorities. In the following decades they would attempt to play a larger role.

42. The following is based on Machado, Loureiro, Luz & Muricy 1978. An important source for the last part of the nineteenth century is the *Anais da Academia Imperial de Medicina*, published in Rio de Janeiro between 1870 and 1890 and later retitled *Anais da Academia de Medicina*.

In 1839 a dissertation entitled "Medicine Contributes to the Improvement of Morals and the Maintenance of Good Customs" had already spelled out this broad view in all its details. The medical profession, which knows people and the disturbances disorder creates in their bodies, should lead the organization of society, finding the causes of social illnesses and interfering to redress them. The cure for society's illnesses was to be gained through the avoidance of passions and disorder. In this doctors' republic, order, calm light, and equilibrium would prevail. The role of medicine was to study the impact of government, freedom, slavery, and religious and other social institutions on the people, to identify the functional alterations they create, and to make the appropriate recommendations for equilibrium.[43] The Sociedade de Medicina do Rio de Janeiro would work persistently to bring society under the scientific supervision of the medical profession while fighting all nonestablished forms of medical work, from homeopathy to traditional medicine.

It is probably fair to say that the Brazilian medical profession never had the same power to put forward their ambitious propositions as the engineers. The market for medical private practice was always better than for engineers, and the doctors could adhere more closely, and earlier, to the canons of a liberal profession. Only doctors more related to general hospitals, sanitary medicine, and the military would attempt a broader role. Their greatest achievements came in the beginning of the twentieth century, when specialists in sanitary medicine joined with engineers to reorganize and sanitize the urban space, more specifically Rio de Janeiro. This was also the basis for Brazil's most important scientific institution, the Instituto Manguinhos, which was born under a promise of social redemption that for a while seemed real.

If doctors as an organized group never held too much power, they came much closer to the social sciences than the engineers, and they played an important role in shaping the country's dominant social ideologies. Physical anthropology appeared in Brazil as a branch of legal medicine. At the turn of the century, Nina Rodrigues, from the Bahia school of medicine, worked with biological theories seeking links between physical shape and criminal behavior. This literature led directly to questions about the racial characteristics of the Brazilian population, the problems of racial miscegenation, and degeneration.[44] Explanations for the troubles with Brazilians—laziness, luxury, lack of discipline—moved from the old environmental conceptions to the new biological and presumably more scientific theories.

43. Machado, Loureiro, Luz & Muricy 1978:197–98.
44. Stepan 1984.

4

APEX AND CRISIS OF APPLIED SCIENCE

We can see the developments in Brazilian science, technology, and higher education in the first decades of the twentieth century as the interplay between two polar tendencies, one geared toward applied work and short-term practical results, the other more academic and tuned to the more European notions of scientific roles and academic education. As the old imperial scientific institutions decayed, the first tendency was the easiest to get started and to gain support, leading to the establishment of research centers and institutes in agriculture, applied biology, tropical medicine, geology, and engineering.

The academic component would often emerge as "clandestine" activities within applied research institutions, and it would become institutionalized only with the creation of Brazil's first main universities in the 1930s. The consequence was that scientific work seldom had the climate and space for intellectual stimulation and initiative that is often obtained in contexts endowed with a strong academic component. In this chapter we shall follow the developments and transformation of applied science in bacteriological research and geology and conclude with the beginnings of mathematics and the physical sciences. In the next chapter, we follow the creation of Brazil's first universities. First, however, a broader background on this period is necessary.

From the Old Republic to the 1930 Revolution

In 1889 a bloodless military coup brought to an end the Brazilian imperial regime and the reign of Pedro II, which had lasted for almost fifty years. The Empire was centralized in Rio de Janeiro, supported by the traditional aristocracies in the Northeast, and identified with slave-based agriculture. The Republic proved to be much more decentralized and related to the development of a new agricultural economy based on free labor and European migration to Brazil's southern provinces, now promoted to states. Of those, the state of São Paulo gradually emerged as Brazil's economic hub, thanks to the continuous expansion of coffee plantations, European and Japanese migration, and, later, industries.

The republican period inaugurated in 1889 lasted until 1930. These years became known as the "República do Café com Leite" (the Republic of Coffee and Milk), or the years of the "política dos governadores" (the politics of the governors). Both expressions reflect the extraordinary political clout of regional oligarchies of the coffee-growing state of São Paulo and of the cattle-producing state of Minas Gerais. But they understate the political strength of the military, which toppled the imperial regime and elected more than one republican president; the historical links between the military and the positivist oligarchy that controlled political life in the southernmost state of Rio Grande do Sul; and a growing middle class, imbued with urban values and raising aspirations, which existed in the country's largest cities, most significantly in Rio de Janeiro and São Paulo.

In 1930 the sectors left out of the "política dos governadores" brought the "República do Café com Leite" to an end and inaugurated the fifteen-year period in which Brazil was to be governed by Getúlio Vargas, a direct product of the Rio Grande do Sul oligarchy. From 1930 to 1937 Vargas engaged in a complicated power play with the military, the states' political oligarchies, the Catholic church, the left-wing intellectuals, and the integralistas, the Brazilian fascists. In 1937 Vargas suspended all legal political activities and declared himself dictator under a new constitutional charter that was supposed to inaugurate a new Brazilian state, the "Estado Novo."

The Vargas years are a watershed in Brazilian contemporary history.[1] Power was again concentrated at the national government, and there were systematic attempts to modernize the state administration,[2] to create a nation-wide education system,[3] and to stimulate industrialization.[4]

It is impossible to appreciate these developments without a proper understanding of the growing rift between Brazil's central political authorities and the country's main economic pole, São Paulo.[5] Since the very beginning, the old captaincy of São Vicente (where São Paulo started) developed independently and far from the colonial central administration, which had its seat in Salvador and later Rio de Janeiro. Travelers in the seventeenth century used to describe it as a "republic of bandits." São Vicente was the first settlement that moved from the coast to the hinterland, in open contradiction to the general pattern of settlements along the coast. The history of the expansion of São Vicente is symbolized by the "bandeiras," Indian hunting expeditions that penetrated farther and farther south, resulting in military clashes with the Spanish Jesuit missions; or expeditions in search of gold and gems, with eventual clashes with other immigrants over mining areas, sponsored and stimulated by the crown; and a conspicuous absence of the province of São Paulo from the forefront of national events until the explosion of coffee plantations in the nineteenth century.

Around 1860 some 80 percent of Brazil's coffee production came from the province of Rio de Janeiro; at the turn of the century São Paulo accounted for more than 60 percent of a much larger production. This dramatic shift is explained in large part by the development of a strong entrepreneurial mentality among São Paulo elites, which included a strong effort to open the region to European migration as a replacement for slave labor and to develop an international policy of price supports that became known as "valorization."[6] Meanwhile, the old agricultural elites in Rio de Janeiro and other regions turned from economy to politics as a way to preserve their traditional positions of status and power. The Paulista elites played a very active role in the downfall of the Empire in 1889 and, for the first time in Brazil's history, shared power with other leading states and the military during the First Republic. In 1930 they found themselves on the losing side against the political oligarchies of

---

1. Skidmore 1967; Schwartzman 1982.
2. Schwartzman (ed.) 1983.
3. Schwartzman, Bomeny & Costa 1984.
4. Wirth 1970.
5. Schwartzman 1975 and 1982.
6. Delfim Neto 1959; E. P. Reis 1979.

Rio Grande do Sul, Minas Gerais, the Northeast, and the young military. In 1932 the state was shaken by a frustrated armed attempt to put an end to the interventionist policies of the Vargas regime. In the aftermath, leading members of the Paulista elite were sent to exile in Europe, to return in the conciliatory years of 1933 and 1934, when a new Constitutional Assembly was supposed to bring the country back to political democracy and decentralization. It was precisely in 1934 that the Universidade de São Paulo was created.

The "Brazilian Enlightenment"

Some authors call the final decades of the nineteenth century and first decades of the twentieth the "Brazilian Enlightenment."[7] It was a time of intense contact with Europe, especially France, introducing Brazil to the concepts of evolution, biological and social Darwinism, positivism, and philosophical and political materialism. Brazil's political, cultural, and intellectual elites welcomed these ideas, each group adopting the aspect that suited it best. Positivism reigned in military circles, and the emperor himself was an enthusiastic propagator of new technologies.

We can only begin to analyze how and to what extent Europe influenced Brazil's intellectual, institutional, and political development. Brazil transplanted often distorted versions of intellectual and institutional models from France and Germany, frequently with some delay. Brazil's intellectual elite also flocked to study in these two nations, especially France. Many scientists and researchers who were to head Brazil's research institutions came from these countries. The British culture did not have much influence in Brazil, though Great Britain was Brazil's main economic partner. Economics and culture did not go the same ways.[8]

Because of the central role it assigned to science, rejecting a speculative or contemplative vision of reality, positivism encouraged Brazilians to accept the new techniques and knowledge that had dominated the European intellectual scene for so long. At the same time, positivism brought with it a vision that had little to do with Brazil's particular reality and opposed the manner by which scientific activities developed in Europe.

In France, positivism was accepted by only some of the evolutionist social philosophers; most natural scientists did not follow it at all. In the

---

7. Barros 1959.
8. Manchester 1933; R. Graham 1968; Needell 1987.

social sciences, positivism confronted new tendencies and theories, such as Marxism, Spencerianism, and historicism. Within the physical sciences, positivism conflicted with the theoretical lines followed in physics since Alessandro Volta and Luigi Galvani revealed the existence of non-Newtonian forces after the eighteenth century. Positivism also ran up against a barrier within mathematical sciences, where work had been influenced by the studies on noneuclidean geometry carried out by Carl Gauss, Nicolay Lobachevsky, and Georg Bernhard Riemann by the end of the nineteenth century. Comte and his followers believed that the concepts derived from noneuclidean analyses were abstractions originating from the metaphysical stage of human thought and should not be taught in schools. Almost totally shut out from the academic community, Comte began preaching to lay audiences. Thus was born positivism's religious branch, whose spokesman was Émile Littré.

Religious positivism arrived in Brazil at full strength. Benjamin Constant Botelho de Magalhães, a military man and a founding father of the Brazilian Republic, stated:

> Positivism is a new religion—the most rational, the most philosophical, and the only one that emanates from the laws of nature. It could not have been the first religion because it requires knowledge of nature's laws and is a spontaneous consequence of this knowledge. Therefore it could not have appeared during the childhood of human reason, or even when the sciences were still embryos; it still would not have appeared now were it not for that remarkable genius Auguste Comte, whose vast intelligence allowed him to leap centuries into the future, seizing science in its definite form and giving us, through his scientific religion, mankind's definitive religion.[9]

Science is achieved; the world is understood. There can be no more room for questioning, doubts, or experimentation. What remains is the need to move on to action, proselytizing the nonbelievers. Within this framework, where does one fit the notion of a laboratory, a research center, a university concerned with expanding the boundaries of the unknown?

While, in Brazil, science was seen as done and ready to use, in Europe and the United States the excitement was barely beginning. Culturally isolated from the Anglo-Saxon world, Brazilians followed at a distance

---

9. Letter sent to his wife from the theater of operations of the Paraguayan war in the 1870s, as quoted by Lins 1967:39.

most of what was happening in engineering but saw little of the developments in physics. As a younger witness recalls:

> All of us—including those who studied at the old Escola Politécnica—were strongly influenced by nineteenth- and twentieth-century French physics during our formative years. [In France] such important figures as Poincaré and Madame Curie certainly made enormous contributions. But French physics was crystallized in various manuals and treatises such as the *Ganon Manouvrier,* the *Tourpin,* and other works that dated almost from the beginning of the century and dealt very little with modern physics. Physics as we studied it was meant for engineers: forces, equilibrium, gravity, fluids—in other words, what is known as classical physics and very little of modern physics.[10]

The beginning of this period was marked by the creation of various institutions, mostly in São Paulo, some of which survive today: Campinas' Instituto Agronômico, for agricultural research (1887); the Instituto Vacinogênico, for the development of vaccines (1892); the Instituto Bacteriológico (1893); the Museu Paulista (1893); the Museu Paraense (1894); and the Instituto Butantã, a center for snake venom research and antidote development (1899). In 1900 the Instituto Manguinhos of biomedical research was established in Rio de Janeiro. Except for the Instituto Vacinogênico (which, with the Instituto Bacteriológico, was incorporated to the Instituto Butantã in 1925), these establishments were responsible for much of what was produced scientifically in Brazil until the 1930s.

New higher education institutes also appeared. São Paulo's Escola Politécnica was founded in 1893; the Escola de Engenharia Mackenzie, also located in São Paulo, and Porto Alegre's Escola de Engenharia, were both founded in 1896; São Paulo's Escola Livre de Farmácia and Rio de Janeiro's Escola Superior de Agricultura e Medicina Veterinária both came in 1898; in 1901 the Escola Superior de Agricultura Luiz de Queiroz was founded in the São Paulo city of Piracicaba; and Rio de Janeiro's and São Paulo's Escola de Comércio, both business schools, were founded in 1902.

By 1940 Brazil had ten engineering schools, eleven medical schools, fourteen pharmaceutical and odontological schools, five agronomy and veterinary science schools—and twenty law schools, including both pub-

10. Danon interview.

lic and private but government-monitored schools.[11] A slight trend toward technical fields was visible in the career choices of Brazilians studying abroad. Among those who went to Belgium, for example, a much larger number chose engineering or medicine than any other profession. Belgium had adopted a system of polytechnic institutes along French lines but without the French military and elitist tendencies and with emphasis on practical learning that would facilitate graduates' access to the professional market. Thus, Belgium provided for Brazilians an attractive alternative to the French *grandes écoles,* which were usually not accessible to foreigners.[12]

The scientific institutions created in the first years of the Republic focused primarily on applying their results to meet what were perceived as Brazil's most pressing needs: exploring the country's natural resources, expanding agriculture, and ridding the nation's main ports and cities of disease. These institutions were stimulated by the industrial growth and development then overtaking Brazil with the opening of new transportation routes (mostly railways) and the expansion of new crops. As the nation's economy grew, unexpected obstacles to further expansion and consolidation appeared—for example, agricultural plagues, cattle disease, and endemic diseases that reduced labor's productive capacity and closed the nation's ports to navigation; the lack of an efficient road, port, and rail network; and energy deficiencies. Bubonic plague at the ports of Rio and Santos, attacks by coffee borers, malaria afflicting workers opening new roads—all these problems demanded a concentrated effort to eliminate them. They were dealt with more efficiently than one might have expected from the precarious public administration inherited from the Empire. Within a five-year period the mortality rate

---

11. F. de Azevedo 1963:288. For Fernando de Azevedo, the republican regime "neither contemplated nor opted for radically changing the educational system to promote the intellectual renewal of the cultural and political elites needed within the new democratic institutions. Maintaining its almost purely professional character, higher education in Brazil was not enhanced by the establishment of cultural institutes such as schools of philosophy and letters or of science, which could link theoretical research to teaching. Nor were any efforts made to foster a scientific spirit by establishing new bases for the reorganization and reorientation of secondary education, the foundation on which the superstructure of higher education—whether applied or not, professional or not—usually rests" (Azevedo 1963:626). Written in 1940, these statements reflect the author's participation in the movements for educational reform and the creation of the Universidade de São Paulo in the 1930s.

12. A survey lists 217 Brazilian students at the Université de l'Etat de Gand between 1817 and 1914, of which 183 were majoring in engineering. The total of Brazilian students in Belgium during that period was 613, most of whom were majoring in technical fields (Stols 1974:657).

in the city of São Paulo was reduced by half, at a time of intense demographic growth.[13] As we shall see, the Instituto Manguinhos in Rio de Janeiro was getting similar or better results.

It is against this background of political decentralization, cultural borrowing, and practical urgencies that Brazilian science would enter the twentieth century.

From Traditional Astronomy to Modern Mathematics

Organized scientific research in the mathematical and physical sciences began in Brazil within the Observatório Imperial (Imperial Observatory) in Rio de Janeiro, formally created in 1827 but active only since 1845. Throughout the nineteenth century the observatory was headed by French-born or French-trained scientists who usually taught also at the Escola Politécnica.[14] At first, the observatory was dedicated almost exclusively to astronomic calculations, regulation of chronometers, and meteorological observations. In 1858 and 1865 the observatory organized scientific expeditions to observe solar eclipses, which marked the beginning of collaboration with French scientists. One of them, Emmanuel Liais, observed comets in Brazil beginning in 1858 using photographic equipment. In 1874, as director of the observatory, Liais imported new optical equipment from Paris and began working on two major projects: coming up with a precise map of Brazil and studying the orbits of Venus, Mars, and Mercury.[15]

Research in astronomy at the observatory had little connection with the teaching going on at the Escola Militar in those years, and for the astrono-

13. Stepan 1976:140.
14. As of 1845 the observatory was headed by a lecturer of the Escola Militar, Soulier de Sauvre, and from 1850 through 1870 by members of the military (Antônio Manuel de Melo, former minister of war and general artillery commander during the Paraguayan war, and Curvelo d'Avila, former navy commander). Emmanuel Liais, a member of the French expedition that came to observe a solar eclipse in 1858, was appointed director in 1870. In 1881 he was succeeded by Louis Cruls. Born in Belgium and a student of civil engineering at the University of Gand until 1868 and later of the military school, Cruls became friends with Brazilian students there, finally coming to Brazil, where he participated in the Brazilian map commission from 1874 until 1876. Following Cruls, the observatory's next head was Henrique Morize. Although French-born, Morize had graduated from Rio's Escola Politécnica in industrial engineering, where he then played an important role as physics professor until 1925 (Ribeiro 1955).
15. Morais 1955.

mer and mathematician Lélio Gama part of the blame was to be placed on the literary tone of the lectures. "The teaching of astronomy in those years suffered the charming influence of the works of Camille Flammarion. Flammarion's influence on the astronomy of the nineteenth century brings to mind Auguste Comte's influence over mathematics, but the circumscribing, limiting nature of Comte's work stands in contrast to the highly literary tone of Flammarion's astronomy. They were both fascinating penmen, and a torrent of astronomic amateurism sprang from the pages of Flammarion. The colorful language he used to describe the celestial spectacle in the end encouraged inappropriate didactic directions divorced from scientific reality. The astronomer must not let himself be dazzled by the panorama of outer space but should measure it instead within a physical-mathematical context."[16] The other side of this romanticized view of astronomy was the extremely pragmatic actions of government regarding the observatory. "Astronomy didn't have a place to stay; it fit in nowhere, since it was impossible to define it in terms of public services. For seventy years the observatory fluttered from branch to branch without anybody being able to identify the characteristic by which it could be fitted into any scheme of public activities."[17] Under Morize the situation reached its extreme: the observatory changed its name to Diretoria de Meteorologia e Astronomia (Directory of Meteorology and Astronomy) and was transferred to the Ministry of Agriculture, Industry, and Commerce.[18]

With all its shortcomings, the observatory provided a significant counterpoint to the intellectual climate then prevailing at the Escola Politécnica. Physicist Costa Ribeiro wrote:

> Henrique Morize's published works alone, which are scant, cannot be used to evaluate his important contribution to the history

16. Lélio Gama worked at the Observatório Nacional from 1917 until his retirement in 1977 and was its director between 1951 and 1967. The quotation above is taken from a written statement prepared for this study in 1977. For broader biographical references and other primary sources, see Museu de Astronomia e Ciências Afins 1988.

17. Lélio Gama, written statement for the author.

18. In 1933 the observatory got its old name back, and its headquarters were moved to a neighborhood in Rio de Janeiro, São Cristóvão. The observatory, however, did not keep up with new developments. In both Europe and the United States during the 1920s traditional descriptive and positional astronomy were being replaced by astrophysics. In 1937 Domingos Costa was chosen to oversee construction of a regional astrophysical station in the mountainous area of the state of Rio de Janeiro, but the impending outbreak of World War II kept the Brazilian branch of the German Zeiss Corporation from assuming commercial responsibility for the maintenance of this station, and the project was shelved. Morais 1955:126–42.

of physics research in Brazil; one must take into account his heavy influence on Brazilian students in awakening their curiosity and their interest in the experimental work that had previously been relegated to second place and in convincing the government of the need to create teaching and research laboratories and to reorganize many official services on more scientific bases.[19]

The prevailing style can be gathered from an article written by Licínio Cardoso, who was responsible for the Escola's course of rational mechanics and an outspoken positivist, in the first issue of the *Revista da Escola Politécnica*, published in 1897. In his *Geométrie Analitique* he wrote:

> Auguste Comte offers as an example worthy of study the double set of curves that the great geometer Descartes discovered can be derived from a circle. With his characteristic outstanding proficiency, which fortunately has attained world recognition, our incomparable master succinctly provides us with a clear and positive idea of how those curves are generated—in this book that is perhaps the most formidable compendium available. But as we have stated above, having offered this as an example he would not carry out studies on it.[20]

This was the setting against which Otto de Alencar began publishing his work. He was already a well-known mathematician, and publication of his 1898 article against Comte's mathematics[21] started a protracted debate. According to Amoroso Costa, Alencar's best-known student, "followers of positivism thought his article a sacrilege, and ensuing criticisms were perhaps inspired more by faith than by reasoning, but it was a question of geometry, and his objections were irrefutable."[22] Otto de Alencar became responsible for introducing Rio de Janeiro's Escola Politécnica to the works of Alfred Clebsh, George Salmon, Gabriel Könings, and Gaston Darboux; to the treatises on analysis written by Charles Hermite, Camille Jordan, and Émile Picard; to probability calculus; and to the books of physicist-mathematician Henri Poincaré.

Alencar's main disciple was Manoel Amoroso Costa, who continued his

19. Ribeiro 1955:171.
20. Quoted in Paim 1974:111–12.
21. "Alguns Erros de Matemática na Síntese Subjetiva de Augusto Comte" (Some Errors in the Mathematics and Subjective Synthesis of Auguste Comte) in the *Revista da Escola Politécnica*, reprinted in 1903 by the French journal *L'Enseignement Mathématique* as "Quelques Erreurs de Comte."
22. Costa 1971:71.

work in mathematics and in leading the movement against positivism.[23] Amoroso Costa was joined in this campaign by Lélio Gama (who would become director of the Observatório Nacional in 1952), Teodoro Ramos (who would play an important part in organizing the Universidade de São Paulo), Roberto Marinho de Azevedo (who would later become director of the Faculdade de Ciências at the Universidade do Distrito Federal), and Felipe dos Santos Reis (later professor at the Politécnica). They attacked positivism not only for its mathematical mistakes but also for its understanding of the role science was to play in society. In 1923 Amoroso Costa wrote against the fascination with material progress that led people to ignore "the existence of a superior scientific ideal that is higher than man's ability to build a thousand cars a day or to perform an appendectomy in ten minutes."[24] This clash of views transcended scientific and technical circles and was fought in the newspapers. As late as 1925, in reaction to Roberto Marinho de Azevedo's articles on the theory of relativity on the occasion of Albert Einstein's 6 May 1925 visit to the Brazilian Academy of Sciences, where he lectured on the theory of light, Licínio Cardoso wrote an article entitled "Relatividade Imaginária" (Imaginary Relativity), which generated heated discussions in the pages of *O Jornal*, one of Rio's main newspapers.[25]

The Escola Politécnica de São Paulo, established in 1893, was never too involved in these debates. As in Rio, the goal of São Paulo's Politécnica

23. In 1900, Amoroso Costa at the age of fifteen having completed his humanities studies at the Instituto Henrique Kopke, at the time one of Rio's best high schools, entered the Escola Politécnica. In 1919 he presented a dissertation on binary stars, and in the same year he took over teaching Escola's topography and astronomy section. In 1924 he was appointed head professor of the class of spherical trigonometry, theoretical astronomy, and geodesy. Between 1920 and 1925 Amoroso Costa took three courses at the Faculté de Lettres in Paris: introduction to the philosophy of sciences, given by Abel Rey; theory of knowledge, given by Leon Brunschvig; and the theory of the movement of the moon, given by H. Andoyer. Influenced by the first two, Amoroso Costa began to dedicate himself to the philosophy of mathematics and problems of cosmogony. In 1928, at the age of forty-three, he was killed in a plane crash when participating in a commemoration of Santos Dumont's return from Europe to Rio de Janeiro. Several other leading figures of Rio de Janeiro's scientific community also died on that flight.

24. "Pela Ciência Pura," included in Costa 1971:150–52. Lélio Gama, in his introduction to Costa's book, writes: "Amoroso Costa had the privilege of making us aware that just as there is beauty in art, there is beauty in the philosophy of pure sciences. In short, he made us understand that feelings and intelligence are the two secret lyres from which man extracts the melodies he dedicates to nature" (Gama 1971:29–30).

25. The reason for this visit has been of great interest, since it could mean that Einstein had colleagues in Rio whom he could recognize and with whom he could talk. In reality, however, his visit to Brazil was just a stopover on a trip to Buenos Aires. For physics in Argentina at the time, see Pyenson 1984.

was to produce professional engineers. What little academic research was done there was undertaken by a few self-taught professors and did not reflect institutionalized scientific activity. Some applied work did take place, however. From the beginning, efforts at the Escola Politécnica were related to the construction of railways; close ties were maintained with the firms responsible not only for this activity but also for electrical energy generation and the city's trolley system. The Escola's Laboratório de Resistência de Materiais was used to test equipment and material both for the railway and for electrical energy sectors.[26]

Teodoro Augusto Ramos was the most prominent mathematician of São Paulo's Politécnica.[27] Throughout his studies at Rio's Politécnica, he was the leader of his group of colleagues and perhaps the most distinguished of Amoroso Costa's disciples. In 1918 he defended a thesis on the functions of real variables in which he proposed, in the words of Francisco Mendes de Oliveira Castro,

> to base the theory of functions of real variables on the simple notion of polynomials.... Twentieth-century mathematics reached Brazil through this work.... The work begins with an excellent summary of set theory and of the main results so far achieved in the field of functions of real variables, from Cauchy to Cantor, Borel, Baire, and Lebesgue. Written when Brazil had not yet fully grasped the rigors of modern mathematics, his thesis was undoubtedly the most important contribution Brazilian mathematical research could have made before the creation of São Paulo's Faculdade de Filosofia.[28]

26. The first significant research work at São Paulo's Politécnica was carried on by Francisco Ferreira Ramos, who as professor of physics was already taking X-rays in 1896, only one year after their discovery by Roentgen. He was succeeded in 1897 by the industrial engineer Constantino Rondelli, a graduate of the University of Torino. In 1911 Afonso d'Escragnole Taunay succeeded Rondelli. In 1912 Luís Adolfo Vanderley was appointed professor of physics and began some investigations in applied physics. Working with Geraldo H. de Paula Souza (who had been responsible for the creation of the school's Laboratório de Ensaios Materiais, which became the Instituto de Pesquisas Tecnológicas in 1925), Adolfo Vanderley established the energy value of dozens of different kinds of food, did experiments with vegetal fuels, and carried out some studies on the radioactivity of mineral water springs. See D'Alessandro 1943; Meiller and Silva 1949.

27. Born in São Paulo in 1896, Teodoro Ramos took his final exams at the Ginásio Petrópolis high school in 1911. The following year he entered Rio de Janeiro's Escola Politécnica, and he graduated in civil engineering in 1916. In 1933 he was made responsible for hiring faculty at the Faculdade de Filosofia, Ciências e Letras, at the Universidade de São Paulo. He died in 1936 at the age of forty.

28. F. M. de O. Castro 1955:68.

Teodoro Ramos was appointed substitute professor at São Paulo's Escola Politécnica in 1919. According to F. M. de Oliveira Castro, "With the efforts of Teodoro Ramos the Escola Politécnica de São Paulo became Brazil's heart of modern mathematics."[29]

From Sanitary Medicine to Biomedical Research

During the Second Empire and the first decade of the Republic, Brazilian medicine was mostly clinical and sanitary. Nineteenth-century diagnostic and therapeutic resources were scant. The efforts of hygienists—the epidemiologists of their time—were focused on correlating certain diseases with soil, climate, and other environmental conditions. Physicians were consulted and gave opinions about the physical organization of cities, the opening of roads, the land-fill of marshes, the construction of sewers, and the basic regulations for private residences, schools, hospitals, and lodgings.[30]

Modern bacteriological research and sanitary medicine started in São Paulo, in part, because of impetus provided by educational, scientific, and technological initiatives in that state in the first years of the Republic. An additional factor was the poor conditions in the city of Santos. Santos was becoming Brazil's busiest harbor, but foreign ships often shunned it for sanitary reasons. Yellow fever and other diseases were also rampant among the immigrants who came in great numbers through Santos and provided the needed workers for the state's economic expansion.[31]

The first initiative was the creation of São Paulo's Instituto Vacinogênico (Vaccine Institute) in 1892, which was to produce vaccines to protect the nation against repeated epidemics of smallpox. By that time the public health service in the state of São Paulo had been fully reorganized, including the enforcement of mandatory vaccination and booster

29. F. M. de O. Castro 1955:69.
30. The main source of information concerning this era is the *Anais da Academia Imperial de Medicina*, 1870–90, later retitled the *Anais da Academia de Medicina*. For a comprehensive study of medical research in Brazil at the turn of the century, see Stepan 1976. See also Machado, Loureiro, Luz & Muricy 1978.
31. In the early years of the Republic a private group, the Companhia Docas de Santos, received a one-hundred-year lease to operate the Santos harbor. Its owners, Cândido Gaffrée and Eduardo Guinle, established a foundation, the Fundação Gaffrée-Guinle, which strongly supported most initiatives related to medical research in Brazil in the following decades.

shot programs and the creation of vaccination posts throughout the state.[32] The law that eventually established the institute also provided for the organization of three different laboratories: one for clinical analyses (which even the private sector did not yet boast), one for bacteriology, and a third for pharmaceutical research. Only the second of these became a reality.[33]

The 1893 creation of the Instituto Bacteriológico (Bacteriological Institute) was foreseen in the same legislation that had created the Instituto Vacinogênico. Its task was "to be especially concerned with microscopy and bacteriology in general and their application to the study of the epidemic, endemic, and epizootic diseases that appear in our midst and become increasingly serious."[34] Given the magnitude of the task and the lack of local experience, it was necessary to take a different route from that of the Instituto Vacinogênico and call on the academic and organizational capabilities of a foreigner. Louis Pasteur himself was consulted, and he suggested Felix Le Dantec.[35] Le Dantec remained in Brazil for just four months, returning to France with the materials he had collected for the study of yellow fever. He was replaced by Adolfo Lutz.[36]

In the end it was Lutz, not Le Dantec, who was responsible for setting up Brazil's and Latin America's first modern bacteriological laboratory

32. There was no secret about the method for preparing Jenner's smallpox vaccine, whose technology had remained unchanged (though somewhat perfected) since its invention at the end of the eighteenth century, but Brazil had nonetheless relied on imports before the establishment of the Instituto Vacinogênico.

33. Blount 1971.

34. Amaral 1958:381.

35. Gabriel Pisa, then Brazil's ambassador to France and entrusted with making contact with Pasteur, reported: "In answer to my letter, the illustrious scholar Pasteur has recommended his disciple Felix Le Dantec to head the Instituto Bacteriológico, considering that Mr. Le Dantec is from all aspects worthy of this post, alumnus of the École Normale Supérieure, Doctor of Natural Sciences, and laboratory assistant at the Pasteur Institute" (Campos 1954:518).

36. Born in Rio de Janeiro of Swiss parents, Lutz studied medicine at the University of Bern, from which he graduated in 1877. Afterward he visited several medical centers in Europe, making contact with J. Lister in London and Pasteur in France and working in dermatology with J. Unna in Hamburg. He came back to Brazil in 1881, had his degree validated by Rio de Janeiro's Escola de Medicina, and began to work as a physician. He worked with leprosy patients in a small town in São Paulo's countryside, Limeira, and published several articles on the subject in the *Zeitschrift für Dermatologie*. He is supposed to have been the first researcher to provide the full description of the leprosy bacillus, a primacy that was later obscured by better-known authors. In 1889 he was invited by J. Unna to work in a leprosy hospital in Hawaii. Having returned to Brazil in 1893, he was invited to become the vice-director of the Instituto Bacteriológico, replacing Le Dantec, and he became its formal director in 1895 (Campos 1954:518; Martins 1955:222; Stepan 1976:139–40).

and for introducing the most advanced techniques then available. The laboratory not only identified diseases and pursued other applied studies but also provided support for such routine activities as blood and urine analyses and vaccine and serum production.[37] As early as August 1893, Lutz demonstrated the practical usefulness of his knowledge, identifying in only one day the unknown epidemic then sweeping a São Paulo immigrants' hostel: asiatic cholera.

In 1894 and 1895 the institute responded rapidly and efficiently to cholera epidemics. The Instituto Bacteriológico, and Lutz in particular, were to gain special fame with the public health campaigns aimed at wiping out yellow fever and the bubonic plague. These campaigns provided a testing ground for interaction among scientists, the public administration, and the population and served as a rehearsal for the national campaigns to be proposed and executed by the Instituto Manguinhos a few years later. It was also a chance for the future great names of the biological sciences to meet, collaborate, and exchange experiences.[38]

In spite of its earlier achievements, the Instituto Bacteriológico's activities and prestige began to decline in the early 1900s. Its budget was not increased significantly, and Lutz had to spend much of his time on bureaucratic chores. In 1908 he accepted an invitation from Oswaldo Cruz to join the team of researchers at the Instituto Manguinhos in Rio de Janeiro. Vital Brasil had already left the Instituto Bacteriológico in 1899 to direct work on the production of antibubonic serum at the Butantã ranch on the outskirts of São Paulo; once Lutz also was gone, no one was left to continue scientific research. Although Lutz was still its formal director until 1913, the institute gradually lost its raison d'être as a separate body, and in 1925 it was absorbed by the Instituto Butantã. In 1931 it was revived and reorganized as the Instituto Adolfo Lutz.

The new views on tropical medicine would take almost ten years to travel from São Paulo to the capital city of Rio de Janeiro. In 1897 Brazil's director of public health, Nuno de Andrade, addressed a memorandum to the Academia de Medicina inquiring about the advantages of

37. Stepan 1976:140.
38. When Lutz identified the fever afflicting São Paulo in 1895 as typhoid, he had to contend with the opposition of the newly created Sociedade Médica e Cirúrgica de São Paulo, which refused to accept a diagnostic methodology based on the identification of causal organisms. They insisted on the traditional view that epidemics were caused by environmental conditions such as the weather, a notion that led to the very concept of "tropical diseases." The impasse was broken democratically by a vote, which Lutz lost. According to him, doctors at that time "systematically oppose all progress, basing their ideas on the works of authors who are either not competent or out of date" (quoted in Stepan 1976:141).

"fostering the establishment of official technical institutes to prepare antitoxic and healing serums." He also asked about the advantages of setting up official institutes to prepare serum and vaccines, about the validity of bacteriological research being done in Brazil, and about the advantages of restricting the institute to Brazilian nationals. In response he got the backing of the academy for his undertaking.[39]

The project came to life in 1889. Threatened by the bubonic plague from São Paulo and facing problems in importing serum directly from Europe, Rio de Janeiro's Mayor Cesário Alvim founded the city's Instituto Soroterápico Municipal. Technical control was handed over to Oswaldo Cruz, who after three years of specialization at the Pasteur Institute in Paris had helped identify the bubonic plague epidemic in Santos, in association with Adolfo Lutz and Vital Brasil.

Less than one year later, in May 1900, the institute came under federal control, but its administrative and technical staff was retained. By February 1901 the first one hundred vials of serum were ready. This initial task involved more than the simple reproduction of already known formulas, since the technique used to produce them in Europe was not yet in the public domain, unlike the smallpox vaccine. It was up to Oswaldo Cruz to change or standardize various aspects to achieve a product that would be more efficient, stable, and adequate under Brazilian conditions.

In 1902 Oswaldo Cruz replaced Pedro Afonso as the institute's director. From its initial role as a factory of serum and vaccines, the institute rapidly broadened into a center for bacteriological research and personnel training and a gathering place for a new generation of medical doctors in tune with the medical revolution started by Pasteur: Miguel Couto, Carlos Chagas, Eduardo Rabelo, Marques Lisboa, Cardoso Fontes, Ezequiel Dias, and Artur Neiva. Under the guidance of Oswaldo Cruz these scientists produced excellent results in the fields of hematology, malaria, prophylaxis and etiology of plagues, tuberculosis, infectious diseases, microbiology, medical zoology, insect contamination, and verminous diseases.[40]

With the appearance of yellow fever in 1903, Rio de Janeiro faced a

---

39. Nuno de Andrade was a founder of the Policlínica Geral, then the most important general hospital in Rio de Janeiro, and a pioneer in bacteriology in Brazil. He needed the backing of the Academia de Medicina, which had a tradition of providing advice on controversial questions regarding public or private health. The academy expressed its support by making a favorable judgment concerning the quality of the studies and the bacteriologists themselves, some of whom had already accumulated a significant amount of experience: Francisco Fajardo, Adolfo Lutz, Chapot-Prévost, Virgílio Otoni, Oswaldo Cruz, Batista Lacerda, Ismael da Rocha, Pinto Portela, and Clemente Ferreira (*Anais da Academia de Medicina* 1897:71, 77).

40. Guerra 1940:70; Neiva 1941:70.

new threat, and President Francisco Rodrigues Alves appointed Oswaldo Cruz to substitute for Nuno de Andrade as head of the Diretoria Geral de Saúde Pública. Cruz also remained at his post as head of the Instituto Soroterápico. As a result, sanitary control within Rio de Janeiro and other areas of Brazil could be exercised through the integrated efforts of pure and applied work.

The rise of Oswaldo Cruz to the Diretoria Geral heralded the start of a highly productive period at the institute. The questions then absorbing the energies of scientists in Paris, Berlin, and the United States coincided with Brazil's sanitary needs. After experiments in Cuba had proven Carlos Juan Finlay's theory that only one type of mosquito, the *Aedes aegypti*, could transmit yellow fever, Brazil became the first important ground for testing this and other modern sanitary theories. As with bubonic plague, techniques developed abroad could not be applied directly without being adapted to the specific conditions in Brazil. Furthermore, a well-prepared team, that was convinced enough of its effectiveness to withstand the active opposition of those who contested the scientific validity of Finlay's claims, would be required to enforce these new methods.

Reactions against sanitation campaigns under President Rodrigues Alves were intense and culminated in the 1904 popular revolt against mandatory inoculation against smallpox. These reactions were not merely a consequence of ignorance or prejudice. They were also directed against Mayor Pereira Passos' plans to modernize the city of Rio de Janeiro, "sacrificing and uprooting the population in the poor downtown areas with the intent of transforming the colonial city, cramped narrow streets and totally lacking in hygiene, into a metropolis with all the characteristics of a modern urban center."[41] The poor suffered most:

> Their belongings were thrown out, their houses demolished, rents raised, and they were moved far from their places of work. In other words, their whole way of life was completely disrupted. From this perspective, one cannot view the reactions against mandatory vaccination and against Oswaldo Cruz himself as antiscientific reactions of the lower classes, who were faced with a cultural element unfamiliar to them, although this may even have been part of it.[42]

The backlash gained ample space in the press and was carried over to the Congress. In large part it served as a pretext to oppose the presi-

---

41. Carone 1971:197.
42. Pena 1977.

dency of Rodrigues Alves. Positivist intellectuals provided justification for this reaction. They challenged the validity of the scientific theories then being developed and the usefulness of their therapy. They fought against what they called "sanitary despotism" and the growing power of the established medical profession in all its manifestations.

> We are not just against mandatory vaccination; we are also against mandatory disinfection, this comedy that forces the citizens to inhale noxious gases and spoil their health; we are against the forced isolation and the way people are violently taken from their families and then allowed to die by the moral actions against their bodies.... We are against mandatory notification of illnesses to the sanitary authorities, which breaks the doctor's vows of professional secrecy, offends their dignity, and forces them to accept the official nosography and diagnoses, in a clear attack on their freedom of thinking and professional work.[43]

In the end Oswaldo Cruz became something of a mythical figure. The population was impressed that a Brazilian sanitarist, heading a team of Brazilians, had succeeded in controlling a disease that was viewed as a major obstacle to the nation's progress. The team earned even greater esteem after receiving first prize at the 1907 International Hygiene Exposition held in Berlin, which established its international recognition. The same year, the Instituto Soroterápico Federal became the Instituto de Patologia Experimental de Manguinhos. Originally entrusted solely with the manufacture of serum and vaccines, the institute assumed the character of a research center. Under its new statutes the institute enjoyed "total autonomy in its technical and scientific investigations" and could ask the government to send any of its staff members to various places to study relevant scientific questions. The institute was also to have its own journal, *Memórias*, for distribution among national medical, veterinary, and agricultural schools and for exchange with foreign scientific journals.[44]

In appointing staff to lead the newly organized institute, the group

---

43. From a letter to *O País*, Rio de Janeiro, as quoted by Porto 1987:57. See also Nachman 1977; J. M. Carvalho 1987.

44. According to the new bylaws established in a decree of 12 December 1907, the reformed institute was to study parasitic and infectious diseases that attack humans, animals, and plants, as well as questions concerning hygiene and zoology; it would also prepare therapeutic serum and similar products that could be used in the treatment and prevention of disease. If the scientific work produced there permitted, the institute would become also a veterinary school, covering the fields of animal pathology, hygiene, and therapy. See Barbosa and Barbosa 1909:155–56.

that had been working there since 1901 was ratified: besides Oswaldo Cruz and Henrique Figueiredo Vasconcelos, there were Henrique Rocha Lima (chief of staff), Alcides Godói, Antônio Cardoso Fontes, Carlos Chagas, Artur Neiva, Ezequiel Dias, Henrique Aragão, and José Gomes de Faria—medical doctors trained at the institute itself. Brazil boasted a "school" of experimental medicine comparable to any of Europe's better centers. At the Instituto Oswaldo Cruz, the French and German traditions blended to add clout to the struggle to discredit the view that Brazil's tropical nature doomed it as a country.

With yellow fever finally under control, Brazil faced a new challenge, malaria. Many public and private works had to be interrupted when health risks affected certain locations. The institute was asked to assess sanitary conditions and to come up with a strategy for implementing sanitation measures. Some researchers were sent out to survey the region's ecology, while others remained at the institute to work on investigations that could not be done *in loco*. All the specialists were to get experience in every area in order to avoid conflicts between laboratory scientists and field specialists.[45] Carlos Chagas' observations of living conditions of both mosquitoes and human beings enabled him to formulate his doctrine of household infection, which led to a change in the techniques used to fight this disease. No longer was it considered important to destroy the clouds of mosquitoes that invaded forests and marshlands; efforts were now focused on eradicating the insects immediately after they had bitten infected people—that is, mosquitoes found in homes. In this way the new medicine reestablished its links with the traditional concerns about the environment.

Many inland posts were set up so that sanitary conditions could be surveyed or a specific problem fought. One of these was located in Minas at the end of Brazil's main railway line, where construction on a planned extension had been forced to stop because of the treacherous sanitary conditions. There, in 1907, Carlos Chagas accomplished what is considered a scientific feat even today: through its causal agent he identified a new disease, the American Trypanosomiasis, which later became known as Chagas' disease. This discovery contributed to building the institute's scientific identity because it opened doors to many new areas of study: the morphology and biology of the Trypanosome; its development cycle in vertebrates and in the carrier; the mechanisms of disease transmission; pathogenic processes; the symptoms and the pathological anatomy

---

45. The need to attain a greater understanding of the mosquito, a carrier of malaria, furnished Brazil with its first entomologists: Carlos Chagas, Artur Neiva, Costa Lima, César Pinto, Gomes de Faria, and Antônio Peryassa.

of the diseased individual; epidemiology; the habitat of the carrier and the conditions for its contamination; and the establishment of preventive and therapeutic norms.[46]

The quality of work being carried out at the institute attracted three German scientists to Brazil—Stanislas von Prowasek, Gustav Giemsa, and Johannes Franz Hartmann—who worked in close collaboration with researchers at the institute during 1908 and 1909. Their arrival certified that Brazilian science had attained a high level, and for some time the institute's mystique was sustained by its excellent production. In 1910 while researching malaria, Artur Neiva demonstrated the existence of a type of plasmodium that was resistant to quinine. In 1911 Gaspar Viana identified the *Leishmania brasiliensis,* and in the next year he discovered a treatment using emetic tartar. With Henrique Beaurepaire Aragão he published two important works, a description of the disease transmission by *hematophagous dipterons (phlebotomus),* and a thorough study of the venereal granuloma: clinical description, histopathology, study, and treatment by emetic tartar. The research on protozoology and entomology proceeded intensively. Studies on mycology and helminthology were also carried on, becoming among the most relevant contributions of the institute.

Financial resources for much of this research came not from federal funding of the institute but from what became known as the "verba da manqueira" (manqueira money). In 1908 Alcides Godói and José Gomes de Faria developed a highly efficient vaccine against "manqueira," a disease that afflicted Brazilian cattle. They donated the patent on this vaccine to the institute, and profits from the sale of this product began to equip laboratories, pay new researchers, and finance staff trips around Brazil or to neighboring countries in search of new problems and new solutions.[47]

The donation of this patent tells us something about the climate that prevailed in the institute. Shut away on a farm on what was then at the fringes of Rio de Janeiro, institute scientists saw themselves as a very special group of people dedicating their lives to a cause more noble than most. For this very reason, it was extremely difficult to break into the group. Whoever wished to join the circle first had to be accepted into a very demanding practical course, after completing the first years of medical school. To earn the right to a later internship at the institute, candidates had to have perfect attendance in the two-year course. Students

---

46. Fonseca Filho 1974:46.
47. Neiva 1941:64.

were still on probation during the period of internship, doing unpaid work for the staff researchers who agreed to take them on, until the opportunity arose for the candidate to join the permanent staff. The candidates themselves felt that such tests were necessary to gain admittance to what was then considered the only institution in Brazil where real science existed. Besides providing a stimulating environment, the institute had an excellent library, a good infrastructure, and a fine technical staff, including glass blowers, electricians, and machinists, all trained by the senior researchers themselves. Once admitted, all candidates could expect to have their work not only recognized but also used in many campaigns promoted by the health authorities to which the institute was linked.

Geological Research and Economic Nationalism

The third area of applied research begun at the turn of the century covered geology and mineralogy. A series of short-lived geological and geographical commissions had been established since 1875, headed by American-born geologists and later by graduates from the Escola de Minas de Ouro Preto. The first, the old imperial Comissão Geológica, was reborn in 1907 as the Serviço Geológico e Mineralógico (Geological and Mineralogical Service), a federal agency whose directorship was offered to Orville A. Derby.[48]

48. The Comissão Geológica do Império (Imperial Geological Commission) lasted from 1875 to 1877. It was headed by Charles F. Hartt, who had been to Brazil with the 1865–66 Thayer expedition, under the direction of Louis Agassiz and who in 1871 headed Cornell University's Morgan expedition. In 1870 Hartt published *Geology and Physical Geography of Brazil*, a book based on his earlier travels. Members of the commission included Americans Orville A. Derby, John Caspar Branner, and Richard Rathburn and Brazilians Pacheco Jordão and Francisco J. de Freitas. Derby was invited to organize São Paulo's Comissão Geográfica e Geológica in 1886, where he worked with E. Hussak and two graduates of Ouro Preto, Luís Felipe Gonzaga Campos and Francisco P. Oliveira. He was the first director of the Serviço Geológico e Mineralógico do Brasil, from 1906 until his suicide in 1915. Two other short-lived institutions were established in 1891—the Comissão de Exploração Geográfica e Geológica de Minas Gerais and the Comissão Especial do Planalto Central do Brazil (Special Commission for the Brazilian Central Highlands)—which did the first studies to determine the location of Brazil's future capital, Brasília. This period was also marked by coal research efforts carried out by the Comissão dos Estudos do Carvão (Commission for Coal Studies). Headed by American geologist I. C. White, that commission made stratigraphic surveys of southern Brazil during 1904–5 (Leonardos 1955; Leinz 1955; Pereira 1955).

As the director, Derby had the cooperation of two former associates, Eugène Hussak and Gonzaga Campos, and tried to build the agency in the same research tradition with which he had graced other institutions. In spite of the scientific achievements of this group, however, the new institution did not fare well, and in 1915 Derby committed suicide, attributed by some authors to the government's disregard for the Serviço Geológico. After his death, applied research received greater and greater emphasis: "In this phase of applied geology, preference is given to economic topics— petroleum, hydraulic energy, iron, coal, and even agricultural soil—in addition to the geographic surveying of the Amazon basin and the publication of many maps of different regions of the nation."[49] Derby was succeeded by Gonzaga Campos, a graduate of Ouro Preto's Escola de Minas, who held this post until 1924, when he was replaced by another graduate of the Escola de Minas, Eusébio Paulo de Oliveira.[50]

The Serviço Geológico grew under the jurisdiction of the Ministério da Agricultura, Indústria e Comércio (Ministry of Agriculture, Industry, and Trade). Jesus Soares Pereira, a civil servant who was to play an influential role in Brazil's nationalist economic policies of the following decades, described the place as a very special institution, full of public spirit and dedication.[51] Other observers viewed it much the same way.

49. Pereira 1955:369.
50. From then on, alumni of the Escola de Minas de Ouro Preto dominated the earth sciences: statesman Pandiá Calógeras, author of the classic *As Minas do Brasil—sua Legislação* (1905); Miguel Arrojado Lisboa, considered the most important geologist of his time; and a long list of researchers for the Serviço Geológico including Fleury da Rocha, Alberto Betim Pais Leme, Avelino Inácio de Oliveira, Paulino Franco de Carvalho, José Ferreira de Andrade Jr., Pedro de Moura, Glycon de Paiva Teixeira, Irnack Carvalho do Amaral, Alvaro de Paiva Abreu, and many others. Alumni of Rio de Janeiro's Escola Politécnica also made significant contributions to the geosciences: Othon Leonardos, Ferdinand Laboriau Filho, Sílvio Froes Abreu, and Mário da Silva Pinto, among others. Biographies of these geologists can be found in Leonardos 1955:270–86. From 1927, when he was still an engineering student, Mário da Silva Pinto remembers Eusébio de Oliveira as the man who trained many of Brazil's earth scientists. Under his guidance, Silva Pinto served in all sections of the Serviço (chemistry, physicochemistry, topography, drilling, geology), acquiring a vast general background (Pinto interview).
51. "Within the Ministry of Agriculture—especially in Brazil's former Serviço Geológico e Mineralógico, later transformed into the Departamento Nacional da Produção Mineral—there was a highly enlightened and active core of nationalists. I worked alongside men like Adosindo Magalhães de Oliveira, an engineer who you do not hear much about but a man of high moral stature, the grandson of Benjamin Constant, and one of the pioneers in applying nationalist ideas of natural resources and electrical energy. Many years later he became a director of the Companhia Hidroelétrica de São Francisco." Another key figure was Mário Barbosa Carneiro: "Considered Brazil's top civil servant at his time, he was a man of finest moral conduct and extremely dedicated. He left the Ministry of the Navy to organize the Ministry of Agriculture." It was thanks to Barbosa Carneiro

When Gonzaga Campos was still head of the Serviço Geológico, daily afternoon meetings in the director's office addressed major problems concerning sea transportation, ports, railroads, highways (the execution of vast public works had already begun in the Northeast), fuels, water energy resources, electrical energy, dams, ores, and manufacturing industries.[52] The minister himself, then Hildefonso Simões Lopes, would show up at these gatherings from time to time to join in the discussions.

The two major topics were steel and petroleum. The Brazilian government contracted with American entrepreneur Percival Farquhar and granted him monopoly over the export of ores in exchange for the construction of a steel plant in Brazil. The contract had generated much debate ever since it was signed in 1920, which dramatized a debate that would be present in Brazilian economic life in the decades to follow. The liberals argued for opening up the country to foreign ventures and accepting its role of a supplier of agricultural products and raw materials to the industrialized centers, while the nationalists strived to encourage domestic industrialization through public incentives and the establishment of state control over natural riches.[53] There was a clash of ideas but also of regions and groups. Scientists and technologists viewed their roles in Brazil's future economic growth differently. The nationalists, mostly graduates from the Ouro Preto school, tended to see themselves as civil servants responsible for leading the country on the road of progress. The liberals were mostly from the Escola Politécnica in Rio de Janeiro and usually combined their technical role with entrepreneurial activities, either as contractors for the state or in association with large Brazilian or international economic groups.

In 1921 the Ministry of Agriculture created the Estação Experimental de Combustíveis e Minérios in Rio de Janeiro, which was to become Brazil's first technological research institution in the modern sense, with the purpose of continuing and broadening studies of the energy potential of coal deposits in southern Brazil. Soon other fuels and mineral resources were also included.[54] As the station's first director, Ernesto Luís da Fonseca Costa sought to attract the most qualified personnel to his team, among them Sílvio Froes Abreu, his favorite disciple

---

that Jesus Soares Pereira entered the ministry and later became part of the Departamento Nacional da Produção Mineral at the time of its 1934 creation. Both Carneiro and Magalhães de Oliveira were avowed positivists (Pereira 1975:38, 58).

52. Rosa 1974:2.
53. Wirth 1970: part 2.
54. Schwartzman 1983; Schwartzman and M. H. M. Castro 1984.

and successor.[55] Concern over the nation's energy resources shortly spurred technological research into the use of alcohol in combustion engines. In response to a sugar glut and the lowered competitiveness of this commodity on the international market, the Brazilian government decreed in 1931 that alcohol be mixed with gasoline at the pumps in a concentration of 5 percent.[56]

As time passed, the station broadened its range of activities and thereby attracted a growing number of researchers, mostly from Rio's Escola Politécnica. In 1933 it came under the jurisdiction of the Ministry of Agriculture's short-lived Diretoria Geral de Pesquisas Científicas, headed by Juarez Távora. One year later the station came under the control of the Ministry of Labor, Commerce, and Industry, created at the end of 1930, and received the name it bears today: Instituto Nacional de Tecnologia. The new institute maintained the routine technological work and goals of the original station, and new fields of work were added: metalmaking, construction materials, physics and chemistry, electricity, fermentation, and others. Equipped with excellent laboratories for its day, the Instituto Nacional de Tecnologia became for some time Brazil's main center for technological research activities, excluding biomedical sciences. In 1934, while still director of the Instituto, Fonseca Costa brought in a young German researcher and engineer, Bernard Gross, who was to become the institute's leading resident scientist.[57]

Behind the transition from the Ministry of Agriculture to the Ministry of Labor was the institute's leadership opposition to the nationalist orientation that was to prevail under Juarez Távora. Fonseca Costa was strong

---

55. Other staff members included engineers Paulo Accioly de Sá; Aníbal Pinto de Souza; Britain's Thomas Legall, a specialist in ovens and coal combustion; and Heraldo de Souza Matos, who supervised research on the use of ethanol in spontaneous combustion engines and was later put in charge of the division of thermic fuels and engines. Industrial chemists Joaquim Correia de Seixas and Rubem de Carvalho Roquete were also part of the team.

56. This was made possible through the successful research results. To show the possibilities, an alcohol-fueled car made experimental trips between Rio and São Paulo and between Rio and the neighboring mountain city of Petrópolis, and in 1925 a team from the station participated successfully in a car competition to demonstrate the technical feasibility of its proposals. In the 1970s, as a response to the oil crisis, Brazil engaged in a full-scale program of replacing gasoline with alcohol, a project that counted on the technical participation of the same institution, now called the Instituto Nacional de Tecnologia.

57. Gross had arrived in Rio a year earlier without any firm professional goals. As a newly trained researcher in Germany, he had participated in the measurement of cosmic rays. During his first year in Brazil he presented papers at conferences and published articles on this topic, including an article for the Politécnica's engineering journal. In spite of the quality of Gross' work, the institute entered a period of profound deterioration after World War II, from which it would never fully recover.

enough as a personality to keep the institute going and even expand its activities and role during World War II. This was not true of Sílvio Froes Abreu, who replaced him after his death in 1947.

The old Serviço Geológico was on the other side of the fence. In 1934 it was transformed into a new and refurbished Departamento Nacional da Produção Mineral. The new structure included a waters service, a section for the promotion of mineral production, and a central laboratory—besides the geological service, which was responsible for research in geology and paleontology. In addition to its research tasks, the department was to establish norms to execute ore and petroleum policies then taking shape.

The creation of the department was timed with the promulgation of the Mining Code, which for the first time in Brazil's history determined that underground riches belonged to the nation, not to land owners, and that their exploitation would depend on government approval. The new department was created within a tense climate of criticism of the old Serviço Geológico.[58] It was against this background that the new department was entrusted to the direction of Fleury da Rocha, a graduate of Ouro Preto's Escola de Minas.

The tone of the debate can be seen in the role of Monteiro Lobato, best known today as Brazil's leading writer of children's books. Lobato was also a frustrated entrepreneur and indignant about the obstacles the department put in the way of his efforts to find oil through his private company. He was convinced that the department had developed an association with the large American oil companies to prevent Brazil from producing oil, and he looked for German partners to compensate. Jesus Soares Pereira, a long-term supporter of the department's nationalist policies, agreed with Lobato on many counts, but he supported the department's stand as a defense of national resources against predatory foreign exploitation.[59]

58. Sílvio Froes Abreu described the situation: "Private enterprise interested in the mineral industry—especially private foreign entrepreneurs—did not think kindly of this federal body; researchers were leery of the service and, thanks to the ideas planted by Clodomiro de Oliveira, a certain xenophobia among official geologists could be discerned; dissatisfaction with the director, Eusébio de Oliveira, spread as a consequence of the campaign launched by São Paulo and Alagoas petroleum companies" (Abreu 1975:27). Froes himself was far from a neutral observer; while working in association with the Guinle group, he had been surveying the existence of oil in the state of Bahia and planned to create his own oil company, a project that was frustrated by the 1934 Code of Mines.

59. Lobato "accused the government of not being capable of discovering petroleum. To a certain extent this was not surprising. The Ministry of Agriculture's available equipment was faulty. The problem involved not just a lack of funds but how to manage these funds. This kind of criticism was undoubtedly justified." What the government did not agree with

The issue had an unavoidable scientific dimension. Did Brazil have any petroleum or not? The department argued that there was no petroleum located within Brazilian territory, based on the opinion of two specialists contracted from the United States, Victor Oppenheim and Mark C. Malamphy.[60] Petroleum was eventually found, but never in the amounts imagined by Lobato.

The debate was made more difficult because scientific training, as provided by the Escola de Minas, did not enable the geologists of the new department to undertake top-quality geological research. When Viktor Leinz arrived from Germany in 1934 to join the department's newly created petrography section at the invitation of Djalma Guimarães, he found a stimulating but not very professional climate. The Escola's library was deficient and exclusively dedicated to French works, ignoring all German and English texts. Leinz characterizes the Escola de Ouro Preto at the time as "polyvalent":

> It trained engineers of all kinds. Geology of course represented only one small facet of these teachings, and so the geological part was small. There were civil engineers, mining engineers, metal-making engineers. It was evident they had little to offer to the field of geology. Only a few could overcome this problem, amateurishly—that is, through self-teaching. Their colleagues lacked an adequate geological background. . . . They were familiar with Brazil but did not know much about general geological problems. This perhaps is better today.[61]

One way to improve the department's scientific level was to hire foreign scientists, among which Viktor Leinz can be considered an outstanding example. The war in Europe provided many others, including world-renowned chemist Fritz Feigl and physicist-chemist Hans Zocher. "At

---

was Lobato's solution to the problem, which was to open the country's resources to private interests. The dominant view at the department was that "the government had to face Brazil's petroleum problem on a scale adequate to its available means" (Pereira 1975:35), which meant that if the government could not extract and control the oil industry in Brazil, then nobody should.

60. Lobato questioned Oppenheim's and Malamphy's qualifications on ethical terms (they supposedly had offered international consultant services on Brazilian petroleum while under contract to the department) but principally on professional terms. Lobato challenged Oppenheim's theses by referring to the work of another geologist, Chester Washburne, hired earlier by the state of São Paulo, who had raised serious doubts about the scientific validity of Oppenheim's work (Lobato 1936).

61. Leinz interview.

one point the laboratory alone had twelve top-rate foreign specialists," recalled Mário da Silva Pinto in an interview. "Men—professors from universities in Austria, Czechoslovakia, Yugoslavia, and Germany—who had many followers and left behind them dozens if not hundreds of contributions, including some with practical utility." But this influx of foreign talent was not enough to transform the department into the basis for an autonomous tradition of scientific work.[62]

São Paulo Takes the Leadership

To a large extent, success in applied science was a main cause of the crisis that pervaded most Brazilian scientific and technological institutions in the 1920s and 1930s and led first to the progressive concentration of competence in the state of São Paulo and later to the creation of Brazil's first higher education institutions with significant research functions. Applied scientific efforts gained support owing to their spectacular achievements, but the price of this support was an image that was difficult to maintain: that almost everything could be solved through science and that scientists therefore deserved wholehearted support. This kind of image was difficult to reconcile with the maintenance of scientific activities over an extended period—only sporadically producing results with more obvious social and economic applications—or with the idea that only scientists themselves can judge the importance of their work.

The Instituto Oswaldo Cruz, or Manguinhos, is perhaps the best example of what was then occurring to a greater or lesser extent in other institutes. After making its initial impact, the institute managed to maintain its prestigious position thanks to a highly talented staff, its ties to the international scientific community, and the administrative and financial autonomy that were guaranteed in its bylaws and sales of vaccines. But

---

62. During World War II, an agreement was signed by the Department, the U.S. Bureau of Mines, and the U.S. Geological Survey through which American geologists came to help map out Brazil's strategic natural resources in a reenactment of the old Comissão Geológica Imperial. This cooperation lasted for about twenty years. In 1953, with Getúlio Vargas again in government, a law establishing the state monopoly over oil production and refinery was passed, and a state-owned company, Petrobrás, was created to that end. This was a direct consequence of the ideas generated at the Departamento da Produção Mineral in the 1930s. In time, Petrobrás launched its own training and research facilities and was influential in establishing the graduate program in engineering of the Universidade Federal do Rio de Janeiro (known by the acronym COPPE), the largest in the country (Nunes, Souza & Schwartzman 1982).

after its initial surge forward, Manguinhos failed to grow or to renew itself; it was equally unsuccessful in preserving its former high standards regarding the scientific work being produced. Salaries were low, financial autonomy was restricted by bureaucratic centralization, and strict criteria of competence for the admittance of personnel began to be abandoned. As the institute lost its high visibility and failed to renew, internal feuds, some of them along doctrinal lines, grew in importance. Cardoso Fontes, an avowed positivist, held divergent views about the nature of transmissible diseases and confronted the group headed by Cruz and Chagas on those grounds.[63]

The law forbidding civil servants from holding more than one post led several of Manguinhos' most important collaborators to resign. Adding to these obstacles was the loss of financial autonomy. In the late 1930s all civil service in Brazil came under the centralized authority of a single agency, the Departamento de Administração do Serviço Público (Department for the Administration of Civil Service) and the institute was treated as a bureaucratic office like all others. In the end, Manguinhos failed to keep step with changes in the handling of epidemics introduced by chemotherapy in the 1930s. It soon lost its status as Brazil's most brilliant center for sanitary medicine.

São Paulo, fast becoming Brazil's main economic hub, succeeded in attracting many talented individuals that Rio—Manguinhos in particular—failed to retain. Three institutions in São Paulo—the Instituto Biológico, the Instituto Butantã, and the Faculdade de Medicina—supplemented this brain drain with their policies of actively hiring talent from abroad and other forms of international cooperation.

63. "During Chagas' final years as director and the first few years of his successor, Cardoso Fontes, scientists who were not very well qualified were admitted to the staff at Manguinhos. Such admissions resulted from personal ties. I [Herman Lent] was witness to the beginnings of a confrontation, of the building of a wall between two groups: on the one side stood those who did nothing even though there was so much to be done, and on the other side stood those who produced, published, worked, and struggled very hard for the funds they wanted.... I believe this was the beginning of an internal struggle and of troubles that later became more complex for the same reason: people on the one hand in greater and greater need of funds, facing greater and greater difficulties, while others, who could have been producing, did not do so with the needed intensity, with the intensity of the first group—and those were the ones to have access to funds, travel, conveniences, other possibilities, and even second jobs outside the institute. The institute no longer had its former spirit as a full-time work center" (Lent interview). Another source of antagonisms involved distribution of the profits from the manqueira and other vaccines. The group unofficially headed by Cardoso Fontes defended distributing these earnings equally among all Manguinhos researchers and scientists instead of using them for the institute's work.

Otto Bier, José Reis, Martins Penha, and others who started in Manguinhos were recruited to the Instituto Biológico in São Paulo by Artur Neiva and Rocha Lima, also natives of Rio. They were later joined by Maurício Rocha e Silva.[64] Otto Bier confirms that

> the Instituto Biológico recruited bacteriologists and immunologists through consultations with the Instituto Oswaldo Cruz, which in the end was the source of scientists . . . who came to fill the first posts at São Paulo's sister institute. The Instituto Oswaldo Cruz told São Paulo's Instituto Biológico who were the best students in its training course in the last three years. This is how Adolfo Martins Penha, José Reis, and I were appointed to posts as bacteriologists and immunologists at São Paulo's Instituto Biológico.[65]

The same pattern was followed by another institution in São Paulo, the Instituto Butantã, which grew out of a laboratory that Adolfo Lutz organized to produce a vaccine against the bubonic plague. Under Vital Brasil since 1901, Butantã began to take on the nature of a center for advanced research into such little-known areas as diphtheria, tetanus, and snake and scorpion antidotes.[66] Afrânio do Amaral, a young scientist trained in Bahia with Pirajá da Silva, was nominated as the new director and in that capacity in 1921 left for a long period in the United States. Amaral worked in the organization of the Antivenom Institute of America in the United States, and before his return in 1927 he was replaced at the Butantã by Rudolph Klauss (former director of Buenos Aires' Bacteriological Institute) and Vital Brasil again.

Upon his return to the Instituto Butantã, Afrânio do Amaral set up a

---

64. "Things had come to a standstill in Rio when I [Rocha e Silva] graduated [1934–35]. . . . Things were very difficult for someone wanting to begin a scientific career. The only possibility was to join Manguinhos, earning starvation wages (if one managed to earn any salary at all) or doing unpaid training work. The wealthy could afford to do this, and they stayed" (Silva interview).

65. Bier interview.

66. Vital Brasil was educated in Brazil, and it was his studies on antivenins that first led him to travel abroad, in 1904, when he was already head of the Instituto Butantã. He had intimate connections with the Manguinhos group, and his stay as head of Butantã can be seen as evidence that scientific activities were firmly established in São Paulo, already independent of Adolfo Lutz' pioneering work. Vital Brasil remained director until 1919, when he was replaced briefly by João Florêncio Gomes, who had been trained at the Manguinhos Institute in Rio de Janeiro. Following a disagreement with the state of São Paulo's sanitary service that same year, Vital Brasil and many of Butantã's other scientists transferred to Niterói, across the Guanabara Bay from Rio de Janeiro, where they founded what is now known as the Instituto Vital Brasil.

new area of specialization: the biochemistry of venom. His second term (1927–38) was marked by his efforts to transplant American academic organization and the German scientific tradition.[67] The institute opened up several new sections during this period, including those of experimental physicochemistry; experimental chemistry; experimental genetics with cytoembryology; experimental physiopathology with endocrinology and pharmacobiology; experimental immunology with serum therapy; virus and virus therapy; medical botany with pharmacognosies (aimed at the cultivation and study of Brazilian medicinal plants); and the traditional departments of ophidiology and medical zoology, bacteriology, and bacteriotherapy; immunology; and serum therapy, protozoology, and parasitology.[68] Besides the Germans, other scientists were brought to Butantã. Some, such as Tales Martins and Lemos Monteiro, came directly from Manguinhos or had spent time there. When the Universidade de São Paulo was created in 1934, the Instituto Butantã was appended as an associated institute.[69]

Since its founding in 1913, São Paulo's Faculdade de Medicina relied heavily on the help of foreign professors, including parasitologist Emilio Brumpt and Italian anatomist Alfonso Bovero. Arnaldo Vieira de Carvalho, founder of this Faculdade and its first director until his death in 1920, was the instigator of this reliance on foreign professors, a preference that was to be felt even more sharply when the Universidade de São Paulo was being established.[70] Ribeiro do Vale recalls that in the 1920s he

---

67. Amaral felt comfortable in both. This is how he spoke of his American experience: "The climate I found at Harvard was very similar—although more ample, as it was richer—to that which I had left at Oxford, where I had spent some time earlier. . . . I could see that at Harvard I would come in contact with what I needed most urgently to study up closely. . . . In the United States, biochemistry was developed by great American scholars who knew German and studied German textbooks, as did I. They took specialization courses in Germany, where it had been proven that research is the basis of progress and that economic achievements depend on research. . . . The nations that did not follow its lead have repeatedly been met with defeat" (Amaral interview). The German presence was evident in the list of scientists Amaral brought from abroad to work at the institute: Karl Heinrich Slotta, Gerhard Szyska, Klauss A. Neisser (experimental chemistry), Gertrud von Ubisch (experimental genetics), and Dionisius von Klobusitzky and Paul König (experimental physicochemistry).

68. Amaral 1958:387.

69. Amaral's ambitions would not be fulfilled for a long time. In the late 1930s, political interference at the institute increased and foreign scientists were pushed out, isolated, or simply became so discouraged by the lack of any research climate that they decided to leave. Directors came and went, with more than twenty serving between 1938 and 1954.

70. Born into a prestigious family, Carvalho had graduated from Rio's Faculdade de Medicina in 1889, and in 1897, at the age of thirty, he had become director of the clinical

"chose the Faculdade in São Paulo, then relatively young and not very much preferred.... Future doctors usually chose to attend the university in Rio, even students from São Paulo. Students from Minas Gerais, for example, also tended to enroll at Rio, ... because they wanted a chance to study under the illustrious Miguel Couto and other great names in Brazilian medicine."[71]

Brazilian science thus found itself facing a paradox. Rio de Janeiro offered a limited but prestigious scientific environment, with places where the large philosophical, economic, and political questions were aired and disputed. São Paulo, in contrast, was much more provincial, a place where things were just getting started and with little visibility and recognition. But the region's wealth meant that the best job offers for researchers were at its institutes. Rio was also witnessing the birth of an ideology that placed a high value on scientific endeavors, on the university, and on twentieth-century rationality, which developed independently and without any direct relation to professional scientific work in the strict sense. The Manguinhos group had an important role to play in this new climate, but even more important was the Escola Politécnica, the starting point and motivating force behind Brazil's cultural and intellectual scene during the 1920s and 1930s. Combined, they had a crucial role in the broad movement for a true university in Brazil—a project that surprisingly would become a reality only in São Paulo.

The concentration of institutional and financial resources in São Paulo and Rio de Janeiro inhibited similar projects in other regions. The best students in Bahia, the Northeast, Minas Gerais, and Rio Grande do Sul— and those who could afford it—would seek Rio and São Paulo for their studies and usually not return to their states of origin. This concentration, however, was never absolute, and an exception occurred in one significant regional pole: Minas Gerais.

The intellectual tradition of Minas Gerais goes back to the state's eighteenth-century gold period, when affluent families sent their children to Europe to study.[72] Even with the end of the gold era, Minas was to remain an important center in terms of Brazilian population, culture, and politics. Up to the end of that century, the state's leadership role was eclipsed only by the presence of the court itself in Rio de Janeiro.

---

staff at that city's public hospital, the Santa Casa de Misericórdia. Much before this time Carvalho had earned a position of prestige and leadership in São Paulo, particularly because of his work as head of the Instituto Vacinogênico upon its creation in 1892.

71. Vale interview.
72. Frieiro 1982.

Although the decline of gold-mining activities and the imposition of an economy of subsistence forced a movement toward rural areas, the state's population was from its beginnings largely urban, and the urban elite did its best to foster the development of culture. The Escola de Minas in Ouro Preto was originally established to train miners, but during the republican period under the protection and in the interest of the state government, it slowly became a professional engineering school. The coming of the Republic also saw the creation of a law school (1892) and the medical and engineering schools (1911). Some prestigious high schools were also founded, both Catholic (such as the Colégio Arnaldo) and public (such as the Liceu de Ouro Preto and later the Ginásio Mineiro, which followed the model of the prestigious Colégio Pedro II in Rio de Janeiro).

Besides, Belo Horizonte attracted people from Rio de Janeiro for an unlikely reason: relief from tuberculosis. Marques Lisboa, Borges da Costa, Almeida Cunha, Hugo Werneck, Ezequiel Dias—all physicians and Manguinhos graduates—suffered from tuberculosis and traded the humid and unhealthy Rio for the mountains of Minas, taking with them their educational backgrounds, their work experience, and their contacts. Ezequiel Dias, for example, was a close relative of Oswaldo Cruz by marriage, and the opening of a Manguinhos branch in Minas seems to have been chiefly a way to prolong his life without interrupting his research career. Belo Horizonte's Faculdade de Medicina was to benefit from the experience brought from Rio by this group.[73] The Rio group of scientists kept it running, joined by J. Baeta Viana, who is recognized for his goiter studies and founder of a local research line in the field of physiological chemistry.[74]

Besides its notable work in developing and producing antidotes for scorpion and snake bites, the Instituto Ezequiel Dias was a veritable

---

73. The Faculdade de Medicina was originally created as a private institute by the physician Cícero Ferreira. Ferreira was related to the Chagas family and came from the same home town, Oliveira.

74. Recipient of a scholarship from the Rockefeller Foundation in 1924, Baeta Viana traveled to the United States, where he worked at Yale and Harvard with Otto Folin and L. B. Mendel, returning to the Universidade de Minas Gerais as its central figure in biochemical research. Another important name in chemistry in Minas Gerais was Francisco de Paula Magalhães Gomes, who after graduating from the high school Liceu de Ouro Preto studied at that city's Escola de Farmácia and went on to graduate in medicine from Rio, where he was a classmate of Oswaldo Cruz. Upon his return to Belo Horizonte, he became the first professor of chemistry at the Faculdade de Medicina, which was noted for its high standards of excellence. Another well-known Minas Gerais personality was Carlos Pinheiro Chagas, the first Brazilian to receive a Rockefeller fellowship in 1915. It is significant that he was also a relative of Carlos Chagas.

intellectual center for Belo Horizonte's academic life. The institute's researchers kept in close contact with Manguinhos, sending many graduates to the Rio organization.[75] The institute had a well-stocked library, and every Thursday major articles were presented and discussed with the participation of Faculdade de Medicina professors not directly linked to the Instituto Ezequiel Dias. At the end of the 1930s the Instituto Ezequiel Dias was taken over by the state government. The idea was to place increased emphasis on industrial aspects of the institute to help finance research activities. A few years later, with the state government under federal intervention by the Getúlio Vargas regime, the appointed governor, Benedito Valadares, decided to turn the institute into a purely industrial establishment to produce antidotes and vaccines, and research activities were prohibited.[76]

Another significant institution was the Instituto de Química that existed within Minas Gerais' Escola de Engenharia.[77] The institute served as a base for the activities of the state and federal government's mineralogical services within Minas.[78]

Alumni of the Escola de Minas were also responsible for the birth of other significant teaching institutions, such as the Escola de Engenharia de Itajubá and today's Universidade Federal de Viçosa, an important center for agricultural studies and research. The schools of law, engineer-

---

75. Staff members at Ezequiel Dias included Aurora Neves, bacteriologist and mycologist; Melo Campos, scorpion and snake specialist; Otávio Magalhães, Ezequiel Dias' successor as head of the institute; and young Amílcar Viana Martins, who joined the institute in 1924 at the age of seventeen.

76. "Valadares appointed a cousin of his to the post of administrative director: Dr. Antônio Valadares Bahia, a totally unknown physician from Papagaio do Pitangui. He used to say that he'd rather split a yard of firewood than reach for a book. As a result, Otávio Magalhães left, and the institute as a research body disappeared" (Martins interview).

77. The Instituto de Química was headed by German-born Alfred Schaeffer, who had received his Ph.D. in chemistry from the University of Munich under the guidance of Adolph von Bayer. Schaeffer's collaboration with Baeta Viana was, to use Leal Prado's word, intense. Despite the dominant American influence on Viana's work, Leal Prado believes it is "possible to say that, albeit remote, the German influence [referring to Schaeffer] exerted over Baeta Viana and even on some of his students (Aníbal Teotônio Batista, Ageo Pinto Sobrinho, and others) helped instill the department with an exacting attitude regarding the instruments and methods used" (L. Prado 1975).

78. Together with the chemists, various mining engineers (most of them Ouro Preto graduates) organized what came to be known as the Laboratório da Rua Bahia no. 52. They included Djalma Guimarães, Otávio Barbosa, Sebastião Virgílio Ferreira, Olinto Vieira Pereira, and Manuel Pimentel de Godói. This group was responsible for the creation of Minas' Instituto de Tecnologia Industrial, founded in 1944 and later transformed into the Centro de Tecnologia do Estado (Instituto de Tecnologia Industrial 1958).

ing, and medicine laid the foundations for the 1927 creation of the Universidade de Minas Gerais.

After that, Minas Gerais remained a place where students could begin their education and even get in touch with people and institutions trying to uphold the standards of scientific work that were being developed in São Paulo and Rio de Janeiro. Most of these students came from a small elite of landed families, and the intermingling of family, intellectual, and scientific ties is impossible to disentangle. Young students would be sent to study medicine or engineering in Belo Horizonte and would often continue their careers in Rio de Janeiro and São Paulo. Minas Gerais' ability to keep these students in their region or to bring new talent from other places was very limited, and the same can be said of other regional centers like Rio Grande do Sul or Recife. More often than not, their academic and research institutions worked mostly as a selecting and breeding ground for the country's central cities.

5

THE 1930 REVOLUTION AND THE NEW UNIVERSITIES

"Educação Nova" and the Catholic Church

The 1889 Republic concluded the formal separation between church and state that was already taking shape in the last decades of the Empire. The new regime gave space to the regional oligarchies that had been kept aside by the Empire, but it did not incorporate in any way the new intellectuals who had begun to appear with the modernization of the cities and the beginnings of industrialization. In the new arrangement, there was no place for those who had fought against the Empire carrying the banners of abolitionism and for the more radical versions of republicanism. The Republic was in many ways less enlightened and less modernizing than the Empire precisely because it yielded so much power to the states and renounced political centralization, a landmark of the imperial times.

It is easy to see how education became a central concern for intellectuals who grew in numbers but were kept alienated by the republican regime. If the country would only recognize the importance of education, the intellectuals—and especially those working on education—would come to the fore in national life. They would then have a chance to use what means they had to solve the problems of backwardness, poverty, ignorance, and lack of public-mindedness that prevailed in Brazil. A new concern for education would produce not only more schools but also more agencies, secretaries, and even a ministry for education—and thus more power and employment for intellectuals.

The propagandists of education in the 1920s were likewise marginal in the republican regime and in their desire for more education, but otherwise they were deeply divided. On one side were those who later became identified as the "pioneers of new education," a group that included Anísio Teixeira, Fernando de Azevedo, Francisco Venâncio Filho, Heitor Lira, Almeida Júnior, Lourenço Filho, and several others. For them it seemed evident that the country's problems would begin to go away when the educational system started to expand, modernize, and be more rational. Fernando de Azevedo describes the conflicts during those days as a struggle between the old and the new, the traditional and modern mentalities—almost a generational conflict. The expression "new education," brought by Anísio Teixeira from his years at Columbia University's Teachers College, had mostly a pedagogical meaning, namely the notion that education should be based on the principles of individual freedom, creativity, and originality of thought instead of on formal teaching and root learning, which prevailed in traditional education. Besides those principles, the 1932 *Manifesto dos Pioneiros da Educação Nova* (Manifest of the Pioneers of New Education) supported the notions of lay education, the creation of a national education system according to norms established by the federal government, and the attribution of a central role to the state in carrying on this task.[1] In other words, the project was to continue and expand the centralizing and state interventionist tradition that had been interrupted by the Republic but that could be revived by the new regime under Vargas. The Catholic church and its more active followers resisted.

Fernando de Azevedo, who had personally followed the road from the traditional seminar to the attempts to bring modernity to the schools, describes the Catholic church in Brazil during the first years of the Republic as going through a crisis of stagnation that was replaced by intense activism after World War I. There was, he said, "a mutual indifference, almost a dissociation, between church and century, between

---

1. The manifest's full text, a broad view of the movement and of one of its leading figures, Fernando de Azevedo, can found in Penna 1987.

religion and the living forces of society." Priestly vocations were extremely rare, and those who chose this road became isolated and did not share the life of other students.[2] What is less clear is how what Azevedo himself described as "the most vigorous Catholic movement of our history, for the breath of its social activism, for a new interpretation of church and century, for the rebirth of religious and national spirit, and for a new fighting mood, not necessarily marked by an ecumenical spirit or by the openness of minds" developed from this state of isolation and lethargy.

This experience of Catholic revival has been the subject of extensive scholarship.[3] One of the revival's main characteristics was the intense militancy of Rio de Janeiro's Cardinal Leme, who would promote dramatic events—such as the inauguration of the Statue of Christ the Redeemer or the consecration of Brazil to Our Lady of Aparecida, both in 1931—gathering large crowds in the streets and pressing the new government to take the church into account in the period of nation-building about to begin.

To this militancy of the official church one should add a new element: the emergence of a small group of Catholic intellectuals gathered around an institute that, significantly enough, took the name Centro Dom Vital under the leadership of Alceu Amoroso Lima, who also used the pen name Tristão de Ataíde in his literary writings. The Catholic lay intellectuals shared with everybody else their dissatisfaction with the country's backwardness, ignorance, and lack of moral fiber and with the corruption and inefficiency of the civilian authorities. Like the others, they believed that the road to national redemption included reconstruction of people through education. Like the others, they also hoped to play an active role in this work of human education and national redemption, and they looked to France for sources of inspiration.

The main difference was that, while some took their inspiration from the French enlightenment and the republicanism of the *dreyfusards*, others found more inspiration in the conservative radicalism of the Action Française. For these, the central values were social order, hierarchy, religious authority, and education guided by religious principles and controlled by the church. The enemies were the ideals of liberalism, individualism, freedom of information and thought, and the power of the state, when not controlled by the church. The agenda had not changed much since the times of Dom Vital in the nineteenth century, and as in those

2. F. de Azevedo 1963:270–71.
3. Todaro 1971; Bruneau 1974; Cava 1976; Alves 1979; Salem 1982.

years this too was a period in which the power and authority of the Roman hierarchy over the universal church was reinstated. The progressive "romanization" of the Catholic church brought Brazilian Catholics closer to Rome than ever before, leading to replenishment of Brazilian parishes with foreign priests and to a search for a much stronger and influential role for the church in social and political matters than the republican constitution had predicted.[4]

The 1930 revolution was received by Catholics with mistrust. The word "revolution" was enough to make shudder those for whom the social order—even the worse—was better than any challenge to authority. Besides, Getúlio Vargas was a product of the positivist political oligarchy of Rio Grande do Sul, and his government would necessarily lead to an increase in political centralization and the strengthening of the state. Soon, however, a political agreement emerged. The state would grant the church privileges in the fields of education, morals, and social order; the church, in turn, would provide the government with social peace and ideological support.

The 1920s and 1930 therefore found Brazil faced not only with new ideas and ways of looking at the world but also with cultural, political, and social movements that were to have extensive consequences in the decades thereafter. In São Paulo the Modern Art Week of 1922 removed the nation's painting and literature from the clutches of archaic classicism, permitting closer contact with Brazilian reality and with Europe's more vibrant experiences in art. In Rio de Janeiro the Academia Brasileira de Ciências was established, and the Associação Brasileira de Educação initiated a movement to enlarge and modernize Brazilian education at all levels.

It would be a mistake to interpret these tendencies as leading to a continuous and uninterrupted trend of social and cultural modernization. In the 1930s, they would be affected by the centralizing tendencies of the Brazilian state; by its profound conservatism, in which a militant Catholic church would play a central role; and by the tensions and contradictions that existed among the central state, regional elites, and a new breed of independent intellectuals. In the following sections, we shall look at the movements toward modern science and expanded education that were taking shape in the 1920s and examine in some detail the experiences of academic institutionalization in Rio de Janeiro and São Paulo.

---

4. Bastide 1951; Cava 1976:11–12.

The Search for Alternatives

Two institutions captured the mood for renovation in Brazilian science and education during the 1920s: the Academia Brasileira de Ciências and the Associação Brasileira de Educação. The academy was established in 1922 as an outgrowth of the Sociedade Brasileira de Ciências, founded in 1916. At the time of its establishment, the Sociedade was linked to the Instituto Franco-Brasileiro de Alta Cultura (the French-Brazilian Institute of High Culture), which had been created under the auspices of the French government, as similar institutes had been created in Buenos Aires and other capitals. Henrique Morize, director of the observatory and professor of experimental physics at the Politécnica, was the Sociedade's first director, holding that post until his death in 1930.[5]

The Sociedade initially held its meetings in the Politécnica's faculty room and was temporarily divided into two main areas: mathematical and physico-chemical sciences. Later came finer divisions: mathematical, physical, chemical, geological, and biological sciences. Publication of its journal, *Revista da Sociedade Brasileira de Ciências*, began in 1917 under the responsibility of Artur Moses.[6] Besides publishing and publicizing scientific work, the academy fostered exchange with foreign scientists, especially French scientists. In 1922 Émile Borel was invited to Brazil to give a lecture entitled "The Theory of Relativity and the Curvature of the Universe." In 1923 there were visits by Emil Gley, Henri Abraham, and Henry Piéron; in 1925 a visit by Albert Einstein on his way to Buenos Aires; and in 1926 visits by Paul Janet, Émile Marchouy, and George Dumas.

The academy essentially played a cultural and intellectual role, acting to foster science more than perform it. It did not, for example, sponsor any research programs of its own. To a certain extent the academy stood as the Politécnica's "antifaculty," a reaction to the delayed penetration of more modern ideas within that school. Perhaps for this reason many of its members were fierce advocates of the new education, research, and teaching principles then being preached by the Associação Brasileira de Educação.

---

5. The first board of directors also included J. C. da Costa Sena and Juliano Moreira (vice-presidents); Alfredo Löfgren (secretary general); Roquete Pinto (first secretary); Amoroso Costa (second secretary); and Alberto Betim Pais Leme (treasurer)—all notable men in the Brazilian scientific milieu (Paim 1982).

6. The title of this journal underwent several changes in the following years: *Revista de Ciências* in 1920; *Revista da Academia Brasileira de Ciências* in 1926; and *Anais da Associação Brasileira de Ciências* as of 1929.

The year 1924 saw the creation of the Associação Brasileira de Educação in Rio. The names of its various departments give us an idea of its goals: primary and teachers' education, secondary education, higher education, professional and artistic education, physical and hygiene education, education in morals and civics, and family cooperation. The association sponsored a broad range of activities, including extension courses, research work, elaboration of draft laws, and—most important—a series of national educational conferences that were to mobilize the Brazilian intellectual and cultural environment after 1927.[7]

Othon Leonardos, a geologist who graduated from Rio's Politécnica and was an active member of the association, recalls that in its first years:

> Every week there were meetings of the board of directors and of the various departments—higher education, secondary teaching, professional teaching. Each department would study a topic to be debated by all. One topic that was studied for many years—and about which the association even put out a publication with interviews with several outstanding professors—was the need for a Brazilian university; another was the need for a Ministry of Education. The association also helped create university extension courses. For example, I was in charge of these courses at the Escola Politécnica, which was more centrally located, in Rio's São Francisco Square. We gave between one hundred and two hundred lectures a year. In the afternoon, cars would pull up at São Francisco Square, most people coming by tramway or bus; things were not as hectic as they are today. Attendance was surprising—the auditorium was usually near full, and curiously enough even sidewalk café waiters would attend these conferences, eager for new learning.[8]

Within the association, those most interested in the question of creating a Brazilian university came mainly from Rio's Escola Politécnica.[9]

---

7. The following is based in large part on Paim 1982.
8. Othon Leonardos interview.
9. "Then Lira—Heitor Lira da Silva, a graduate of the Escola de Engenharia—gathered together his classmates, including Amoroso Costa; Backheuser; Lino Sá Pereira; Ferdinand Laboriau sometime later; the Osório brothers and sisters, especially Alvaro Osório and Branca Osório de Almeida Fialho; a sister to Admiral Alvaro Alberto, Amandina Alvaro Alberto, also a famous teacher, married to Siqueira Mendonça; Júlio Porto Carrero, who brought psychoanalysis to Brazil . . . ; Laura Jacobina Lacombe; Carlos Gregório de Carvalho" (Leonardos interview). The scientists working at Manguinhos did not actually participate in this group, although they maintained closer ties with the Acade-

The association's understanding of education and the Brazilian university is reflected in the many *inquéritos,* or opinion surveys, it had sponsored by the end of the 1920s. These surveys essentially consisted of a series of questions prepared by a panel of specialists and addressed to a wide range of well-known public figures and institutions: the answers were afterward published by the country's leading newspapers or as independent books. In 1927 there were surveys on secondary education and on the Brazilian university question,[10] with the support of the newspapers *O Jornal,* Rio's *Jornal do Comércio,* and *O Estado de São Paulo.* Committee members visited São Paulo, Bahia, and Minas Gerais, and the section responsible for technical and higher education sought the opinion of various specialists and professionals on certain issues, such as what university model could best be adopted in Brazil, whether research institutions should be included within the universities, what teaching methods should be employed, whether state governments should offer the federal government financial assistance, and questions concerning the professional standing of university professors (e.g., salary levels and work hours). The results of this inquiry were published by *O Estado de São Paulo* in 1929.

Also in 1927 a series of national conferences on education were initiated. Amoroso Costa presented a paper on the relationship between the university and scientific research at the first such conference, held in the city of Curitiba, Paraná. During the second conference, held in Belo Horizonte, Minas Gerais, in 1928, Tobias Moscoso presented a paper on the university question. In São Paulo the following year, the third conference, under the presidency of Teodoro Ramos, was dedicated to secondary education, but the problem of defining the university's role also emerged as a central issue.

The recently created Universidade de Minas Gerais decided to present a carefully prepared document in response to the "inquérito." This document affirmed that "Brazilian universities should be granted full economic, didactic, administrative, and disciplinary autonomy and [that]

---

mia de Ciências. A noteworthy member of the Academia was French descendant Henrique Beaurepaire Aragão. "We were all greatly influenced by him; he was a true leader" (Leonardos interview). Laboriau, with Paulo Castro Maia, Tobias Moscoso, and Amoroso Costa, all from the Escola Politécnica, died in the plane crash while going to pay homage to Santos Dumont, who was arriving by boat from Europe in 1928. Leonardos recalls that he had decided the night before to let Amoroso Costa go in his place, as the latter had never flown before.

10. Headed by Domingo Cunha, Roquete Pinto, Ferdinand Laboriau, Inácio de Azevedo, Levi Carneiro, Raul Leitão da Cunha, and Vicente Licínio Cardoso.

their feasibility should be guaranteed by independent funding." This concept of autonomy allowed for differences between universities: "To have only one standardized university for all Brazil would not be recommended. On the contrary, each should be freely organized according to its financial resources and regional, geographical, economic, and social peculiarities, while respecting the supreme national interests." The type of university envisioned would naturally train professionals ("engineers, physicians, jurists, pharmacists, business leaders, agronomists, artists, etc."), but it also would "serve as a center for ongoing scientific elaboration, contributing to the enrichment of collective knowledge, improving our physical well-being, and glorifying humanity's cultural heritage." At the same time, "these must be markedly national—and to a certain extent regional—institutions, reflecting the characteristics of the population that sustains them and lending aid to those special needs of its immediate surroundings."[11]

Three notions stand out from these proposals: the separation of professional teaching and scientific activities, the idea of free research, and the concept of university autonomy. Tobias Moscoso, who presented the theme at the second national conference, expressed the dominant view on the separation of professional teaching and scientific activities clearly:

> It is my belief that when we create universities, we must make a clear distinction between two orientations: . . . the technical and the scientific. The first should result in the development of expertise in applying acquired scientific knowledge to practical, professional life, relying on knowledge of the optimizing economic precepts and processes in general terms as well as applied specifically to our nation. The second is aimed at promoting competence in scientific investigations and in contributions to the advancement of science, . . . all of this directed specifically at Brazilian reality, whenever possible.[12]

These two distinct educational questions were to be clearly reflected at the organizational level: "We need a university equipped with departments of chemistry, physics, mathematics, and the biological sciences, with the full means to carry out scientific research in all branches of pure science; and with departments of philosophy, letters, and the social sciences as well, able to provide a rich cultural background."[13]

11. Campos 1954:80.
12. Quotes in Laboriau, Pinto & Cardoso (eds.) 1929:499.
13. Gilberto Amado, as quoted by Laboriau, Pinto & Cardoso, eds. 1929:354.

The idea that research should be subordinated to the nation's practical needs—or to the demands of professional training—was openly rejected. Alvaro Osório de Almeida put it emphatically:

> The secular experience of peoples whose civilization have advanced or are advancing indicates that preserving the spirit of progress requires not only utilitarian-minded individuals but idealistic spirits as well—idealistic spirits able to satisfy their intellectual needs through the pure contemplation of natural phenomena, through knowledge or study of these, or in the cultivation of letters.... These spirits do not require external excitement or excitement from other individuals in their work. Their work by itself gives the joy and inspiration we all need. They are the source, the origin, of society's intellectual production and progress. As is so well understood by all advanced thinkers, this is why we must maintain these individuals alongside the utilitarian spirits that draw their inspiration from the former, transferring the knowledge they have harvested for adaptation and application to life in human societies.[14]

Tobias Moscoso, meanwhile, argued for university autonomy:

> This project would not necessarily be totally frustrated, but it would most certainly be seriously damaged by state intervention in the administration of these institutions, especially as far as didactic questions are concerned. Because of what I have learned from the lessons of other nations and from our own, I am decidedly in favor of total university autonomy and of full university independence from government and even from legislative branches.[15]

The Francisco Campos Reform

The first officially established Brazilian university was in the southern state of Paraná and did not last. Its birth in 1912 was made possible by short-lived liberal legislation that was replaced in 1915 by the so-called Maximiliano Reform.[16] In 1920 the Universidade do Rio de Janeiro was

14. Laboriau, Pinto & Cardoso (eds.) 1929:354.
15. Laboriau, Pinto & Cardoso (eds.) 1929:168.
16. Cartaxo 1948; Almeida Jr. 1956; J. Furtado 1962; Tobias 1968; Lobo 1969.

created under the new legislation as a merger of the old engineering, medical, and law schools. Neither of these universities amounted to much more than a simple agglomeration of professional schools joined under one fragile rectorate with few attributes. In April 1931 Brazil enacted its first federal legislation[17] outlining the characteristics a university should have. This would later become known as the Francisco Campos reform, in reference to the author of the legislation, then head of the newly created Ministry of Education and Public Health and formerly a contributor to Minas Gerais' educational reforms of basic education.[18]

Science and education were not among the main priorities of the Vargas administration, but Francisco Campos saw that they could not only help legitimize the new regime among the educated but also contribute to the modernization project that was in the minds of so many. His legislation was presented as the outcome of an extended debate and an effort to bring together the extremes. The final text, Campos stressed, was "carefully studied, closely examined, and hotly and broadly debated, with opinions gathered from all streams of thought, from the most radical to the most conservative." Campos continued: "This apparently eclectic spirit, broad-minded and pluralist, does not so much reflect reality as it reflects a desire for this reform to gain legitimacy in the eyes of many different streams of belief during a time of transition." But the stated reasons behind the new project leave no doubt that the intent is to achieve a single monolithic, coherent, and official understanding of a university, in tune with the new regime.[19]

17. The legal documents are three decrees issued by Vargas' provisional government: nos. 19850, 19851, and 19852 of 11 April 1931.

18. Francisco Campos, a legal scholar and politician from Minas Gerais, took an active part in the 1930 revolution and became Brazil's first minister of education in 1931. He would become famous for his sympathy with European fascism and for drafting the 1937 authoritarian constitution of the Estado Novo. He is less well known for his efforts to establish a pact of cooperation between the Vargas regime and the Catholic church. He managed to grant the church the privilege of teaching religion in public schools, among other privileges, and he expected that the church would in return provide the government with inspiration, discipline, and spiritual order. The political pact between church and state would take shape in the Constitutional Assembly of 1934, where the foundations of the new Brazilian society were supposed to be established. A new minister of education, Gustavo Capanema, a former protégé of Campos and an attentive listener to Amoroso Lima, took office that year and would be responsible for a profound reorganization of Brazil's educational institutions.

19. "Although the overall structure is the result of deals and compromises between various tendencies, lines, and preferences, the project has its own individuality and unity; behind this structure lies a reasoning that follows broad, clear, and precise lines in a firm and positive manner, guaranteeing that the plans laid out in its principles of administrative and technical organization are adequately proportioned and balanced" (quoted from the decree's justification, transcribed in Lobo 1969:156–61).

In outlining these objectives, Francisco Campos makes clear his awareness of the ideals concerning university organization that were then popular in Brazil. As he saw it, this university would be

> [an] administrative and teaching institution that unites all higher education under a single intellectual and technical leadership, whether such teaching be utilitarian and professional in nature or purely scientific and with no immediate application, with the double objective of providing the nation's professional elite with technical training while providing a climate propitious for speculative, unbiased talents to pursue that goal that is indispensable for the nation's cultural growth—investigation and pure science.

The project did not expect the double role of professional teaching and research to be carried on in isolation. On the contrary, the university was seen as "an active and militant social unit—that is, a center for contact, collaboration, and cooperation between different desires and aspirations, an intellectual and moral family whose activities are not exhausted within the narrow circle of its own immediate interests; on the contrary, as a living unit it tends to enlarge its area of resonance and influence within the social surroundings, taking on a broad, powerful, and authorized educational role." Thus came need for this university to be organized as an autonomous corporation, "with proposed models of organization appropriate for its internal social life, organizations that would encourage contacts and strengthen ties of solidarity, based on the economic and spiritual interests common to faculty and student body." As a tool for influencing the university's surroundings, extension courses were to be created to distribute "the benefits of the university climate among those not directly connected to university life." To achieve these goals, two conditions would have to be met: university autonomy and the creation of institutions dedicated chiefly to research rather than to professional teaching.

In laying the groundwork for these two conditions, it became clear just how distant the ideal Francisco Campos saw was from the real world. Concerning autonomy, for example, it would be "nonetheless inconvenient and even counterproductive for the teaching system if a sudden and complete break were made with the present by granting universities broad and full teaching and administrative autonomy. Autonomy requires practice, experience, and clear-cut guidelines." The nation's immature university environment presumably did not possess such characteristics. Jealous of its newly acquired power, the state intended to play the role of guardian and educator of the burgeoning university, so that such autonomy would one day become "a conquest of the university spirit—matured, experienced, and equipped with a steady and firm sense of direction and responsibility—rather than coming as a gratuitous and extemporaneous concession that would be more likely to de-educate than to instill universities with a sense of organization, command, and government."

There was thus no practical reality to the idea of autonomy. The first decree within the reform created a permanent body—the Conselho Nacional de Educação (National Educational Council)—to assist the minister. With its members appointed by the president, the council had a broad advisory role and final decision-making powers on many issues. For example, it could endorse "the general guidelines to be applied to primary, secondary, technical, and higher education, above all in response to the needs of the nation's civilization and culture."[20]

The next step was enactment of a statute for the Brazilian universities (decree no. 19851) that gave the minister of education and the Conselho Nacional de Educação authority to approve the internal regulations of any universities established in the country. In 116 articles, the statute defined such matters as the responsibilities of rectors; the organization and roles of university boards ("conselhos"), assemblies, and institutes; the organization of teaching, rights, duties, and promotion rules for professors; admissions procedures; disciplinary rules; and even social activities, including the organization of student unions. The next decree established the Universidade do Rio de Janeiro with 328 articles that covered details ranging from a list of schools within the university to the syllabus for each course each year, ending with a table showing enrollment fees, attendance, certificates, diplomas, and so forth.

In no instance does the Francisco Campos reform acknowledge the possibility of universities organizing themselves in any different manner, eventually competing among themselves to offer the highest quality teaching. The detailed outline of all pedagogic and administrative activities and of all forms of operation, and the need for federal approval for any changes, gave Brazilian universities a rigidity that would be shaken on only a few exceptional occasions.

"Research" as an ideal met the same fate as "autonomy." Francisco Campos' declared empathy with science was an illusion. He placed scientific research with art—an indispensable decoration but certainly postponable. In the project a newly created Faculdade de Educação, Ciências e Letras, would lend the university its truly "university" character, "allowing university life to transcend the limits of purely professional interest, fully encompassing all those lofty and authentic cultural values that lend the university the character and peculiarities that define and distinguish

20. Decree no. 19859, in Lobo 1969:198.

it."[21] In this apparent tribute to the ideal of science as culture, the social and economic benefits to be gained through research, in the short or long run, are ignored. The idea of linking teaching to research is also not recognized. Being a mere ornament, science could wait: "During Brazil's first attempt to establish an institution of higher learning, it cannot be forgotten that peoples such as ours, still undergoing a process of growth and maturation, should not try to organize advanced culture all at once, wholly and exclusively. In order for such an institution to flourish in our midst, it is absolutely indispensable that it yield immediate fruits." Therefore, the Faculdade de Educação, Ciências e Letras, "besides being an institution of advanced culture and pure and unbiased science, should be first an educational institute within which can be found whatever elements are vital to training our teachers, particularly those of primary and secondary levels."[22] In other words, it was to be a teachers' college.

The legislation proposed for law, engineering, and medical courses reflects the varying conceptions of different advisers. The proposed law course was to be strictly professional, beginning with an analysis of economic relations—which "constitute nearly the whole content of law"—and including the study of positive law. More conceptual or speculative subjects such as Roman law or philosophy were to be left for graduate work. The proposal for the engineering course highlighted the need to study theory, emphasizing mathematics, physics, and technological research. The proposed medical course stressed the "technological and scientific organization of medical schools, which makes scientific research original and is an indispensable complement to didactic processes."[23]

In short, the Francisco Campos reform promised a great deal and was greeted by most as a landmark in the history of Brazilian higher education. But it appeared at a time when a strong new regime was coming to power, and it was clearly meant to preempt and paralyze the movement toward a university system based on autonomously organized scientific communities, an idea then being defended by active sectors of the Academia de Ciências and especially the liberal faction of the Associação Brasileira de Educação. Campos' secret dealings with the church, and his fascination with Mihail Maniolescu and European fascism,[24] go a long way in explaining his true intents.

Francisco Campos did not remain in the Ministry of Education for

21. Quoted in Lobo 1969:163.
22. Quoted by Lobo 1969:164.
23. Quoted by Lobo 1969:171.
24. Campos 1940; Schwartzman, Bomeny & Costa 1984:61–66.

long and did not have time to launch his Faculdade de Educação, Ciências e Letras. In spite of its centralizing tendency, the Vargas regime depended very much on regional support; only after 1935 did it start along the road to authoritarian power, which culminated in 1937, when Campos resurfaced as the minister of justice of the Estado Novo. This is why the first two universities in the 1930s were created not by the federal government but by the city government of Rio de Janeiro and the state government of São Paulo. The relevance of the Campos legislation was that Campos' conception of a national university in a centralized system would be adopted again a few years later and lead to the demise of the Universidade do Distrito Federal, the creation of a national Faculdade de Filosofia, Ciências e Letras, and constant tension between the Universidade de São Paulo and the federal authorities. The Universidade de São Paulo survived and became Brazil's leading academic institution; for the rest of the country, Campos set the pattern.

A Liberal Project: The Universidade do Distrito Federal

One of the most significant spin-offs of the movement led by the Associação Brasileira de Educação was creation of the Universidade do Distrito Federal in Rio de Janeiro by a municipal decree.[25] Five schools were to make up the new university: sciences, education, economics and law, philosophy and letters, and arts. The first (and actually only) institution to be created was the Faculdade de Ciências. At its inaugural class, Anísio Teixeira, then secretary of education for the Federal District, summed up the new institution's goals:

> The university has a singular and exclusive role. It must not only convey knowledge; books do as much. It must not only preserve human experience; books do as much. It must not only train tradesmen or professionals in the trades or arts; direct apprenticeship does as much, or at least as much is surely done at more specific schools than universities. . . . What the university must do is maintain an atmosphere of knowledge to prepare the individual who serves and develops this university. It must preserve living not dead knowledge, in books or in the empiricism of nonintellec-

---

25. Decree no. 5513, 4 April 1935.

tualized practice. It must formulate human experience intellectually, doing so with inspiration, enriching and vitalizing knowledge from the past with the seduction, attraction, and impetus of the present.... Knowledge is not an object that is passed down from past generations to ours. Knowledge is an attitude of spirit that slowly comes into being through contact with those who have knowledge.[26]

The new university was marked from the beginning by the intense climate of ideological confrontation that existed among Brazilian intellectuals in those years. The revolutionaries of 1930 had split between the left, which included Rio de Janeiro's mayor Pedro Ernesto,[27] and the much more powerful conservative right, personified by the Catholic church, Francisco Campos, and the military leaders surrounding Getúlio Vargas, among others. At the end of 1935 the Communist party tried to seize power through a military uprising, leading to violent repression and a wave of Communist witch-hunting that included the deposition and arrest of Pedro Ernesto. In July of that year Anísio Teixeira had the inaugural speech for the Universidade do Distrito Federal, already anticipating the years to come.

Secretary of Education Teixeira, born in an aristocratic family from Bahia, was far from a Communist, but the influence of American pragmatism and ideological liberalism during his brief stay at Columbia University was enough to make him a target for Catholic conservatives. His speech began with a defense of freedom of teaching and concluded with images of struggle and death:

> There were those who thought it would be possible to start our university tradition by denying the freedom of teaching which is one of the first conquests of human intelligence in modern times. They believed that there could be a university to enslave, rather than to liberate; to stop life, rather than make it go forward. We all know this reactionary language; it is as old as Methuselah: "The crisis of our time is a moral crisis." "Lack of discipline," "of stability." "We are marching toward chaos." "Toward revolution." "Communism is coming!" They speak like that today, as they spoke five hundred years ago.

26. Quoted in Paim 1982:69–70.
27. On Pedro Ernesto and his role as a forerunner of populist politics in Brazil, see Conniff 1981.

At the end, he dedicated the new university to culture, freedom, and the memory of those who had died fighting for the ideals of freethinking: "All those who have disappeared in this struggle, all those who continue to fight—they make up the large university community that we celebrate with the formal inauguration of our courses. Dedicated to culture and freedom, the Universidade do Distrito Federal is born under a sacred sign that will help it work and struggle for a Brazil of tomorrow that is faithful to the great liberal and humanistic traditions of yesterday's Brazil."[28]

The municipal university was to be created anew, bringing in the best minds, but without the problems of the old professional schools. It was expected, said Anísio Teixeira, that the university would take over the task of molding the nation's intellectual class, a task previously left to "the most abandoned and precarious auto-didacticism," and would finally fulfill the long-standing need for an institution capable of training not only high school teachers but also researchers from various fields. Roberto Marinho de Azevedo, a member of both the Associação Brasileira de Educação and the Academia de Ciências, was made head of the Escola. He managed to bring together a group of teachers who not only were worthy scientists but who also identified fully "with the idea of fostering the unbiased study of the sciences in the hopes of producing researchers as well as good teachers in these fields."[29]

Among those hired for teaching positions at the new Escola de Ciências were mathematicians Lélio Gama and Francisco de Oliveira Castro, physicist Bernard Gross, geologists Djalma Guimarães and Viktor Leinz, and biologists Lauro Travassos and Herman Lent. Others were to join the team later, including physicist Joaquim Costa Ribeiro, then a recent graduate of the Politécnica; chemist Otto Rothe of the Instituto Nacional de Tecnologia; and botanist Karl Arens, formerly assistant to Felix Rawitscher at the Universidade de São Paulo's Faculdade de Filosofia, Ciências e Letras. There were no full-time researchers or facilities for research. Everybody carried on research activities in other institutions, which had the effect of building a bridge between those institutions and the Escola de Ciências. This meant that students paid frequent visits to Lauro Travassos' laboratory at Manguinhos, Leinz' laboratory at the Departamento Nacional de Produção Mineral, and Gross' laboratory at the Instituto Nacional de Tecnologia, where they could observe ongoing research and experiments firsthand.[30]

28. "A Instalação, ontem, dos cursos da Universidade do Distrito Federal," *Correio da Manhã, August 1st, 1935*, quoted in Schwartzman, Bomeny & Costa 1984:211.
29. Paim 1982:84.
30. The flavor of the new school can be gathered by the remembrances of one of its first professors, geologist Viktor Leinz. "The learning experiences shared by these students

In 1936 the school year opened with conferences offered by members of a French university mission, which included Émile Bréhier (philosophy), Eugène Albertini, Henri Hauser, and Henri Tronchon (history), Gaston Léduc (linguist), Pierre Deffontaines (geography), and Robert Garric (literature). Taking the graduation of the first class the following year as a sign that his mission has been victoriously completed, Afonso Pena Jr. turned the rectorship over to biologist Baeta Viana, from Minas Gerais, while Roberto Marinho de Azevedo was to turn the Escola de Ciências over to mathematician Luís de Barros Freire, from Recife.[31]

The first year of the Universidade do Distrito Federal was nonetheless set against an unfavorable background. For one thing, resources for teaching aids, equipment, and work areas were poor. The rectory was installed in the building belonging to the Instituto de Educação (a public teachers school for women), while classes were held both at the Escola Politécnica and at a school located on the Largo do Machado square. Laboratory classes had to be held at the old professional schools of the Universidade do Rio de Janeiro or at the Instituto Nacional de Tecnologia.

More serious was that a frustrated communist uprising in October 1935 prompted the government to turn the country's political climate strongly to the right, and from then on the project was doomed. The Federal District was subject to direct government intervention, and Anísio Teixeira was removed from his office as secretary of education.

---

and young teachers—all about the same age—were as enjoyable as can be. I myself soon introduced a system used in Germany: taking school trips with the students, familiarizing them with the geology of the Federal District up close. We would leave in the morning, go out to Copacabana—where there were still many quarries—or we'd go to the beach. I would explain about the influence of the sea, and we walked around. I also started using slides, which were then a real novelty. I had many slides of geological phenomena printed. We also imported many fossil samples from Germany, material that the students themselves could handle. We taught our students to recognize rocks and minerals using simple but modern methods. In 1937 I went on a long graduation trip with the students to Minas Gerais. Afonso Pena, our rector—and son of [Minas Gerais'] President Pena—gave us a hand. We took this trip to see all Minas Gerais—manganese, iron, the Itabira peak, the gold mine of Morro Velho. Most of these graduates had never been outside the Federal District.... I really pushed practical work, direct handling of material. This is extremely important in preventing learning from becoming 'bookish.' Since funds were adequate, the university allowed us to import whatever we wanted. And import we did, essentially from Germany—projectors, teaching aids, maps, minerals, slides, and microscopes. Books were provided by the university. Against this background we could rapidly establish a place to study geology and mineralogy, a place that offered very good conditions for that time" (Leinz interview).

31. Freire would never complete his move from Recife to Rio de Janeiro.

With his dismissal, many professors left the university, having lost all faith in the project's future.[32]

A Model to Be Followed: The Universidade do Brasil

The Universidade do Distrito Federal was closed because it collided, institutionally and ideologically, with the plans held by the new minister of education, Gustavo Capanema, to build a national university that would conform to the outlines put forward by Francisco Campos a few years earlier—and to fulfill the terms of understanding between the Vargas regime and the Catholic church, which Capanema saw as his responsibility to carry on.

Capanema took over the Ministry of Education in 1934, very likely through the influence of the church, and his personal archives show an extensive correspondence with Alceu Amoroso Lima, through which this leading Catholic intellectual established his influence. In one of these letters, Lima states his displeasure about the creation of the Universidade do Distrito Federal and asks for the dismissal of Anísio Teixeira.[33] In 1937, with Anísio Teixeira out of the way, Capanema placed Lima as the rector of the Universidade do Distrito Federal, apparently to prepare its

---

32. "In 1938 we were still working to build the university, but always fearing that it might be closed. At that time, the time of Capanema [then minister of education], the Faculdade Nacional de Filosofia was being created and—I personally didn't know anything about politics—there was talk that our school would close, that everything would be shifted over to the Nacional. There were these rumors—it'll close; no it won't; it'll be shifted; no it won't. . . . All 1938 was marked by restlessness, and at the beginning of 1939 it became clear the university would indeed close. But nobody was told. Those of us who had contracts naturally thought those contracts would be respected, but they weren't. One fine day, when I was going to pick up my pay, a young woman told me: There isn't any more for the university. In other words, it was never announced officially. Perhaps colleagues of mine who were in closer touch with the political side already knew it. But I only found out right then" (Leinz interview).

33. "The recent establishment of a municipal university, with schools directed by people who made no secret of their Communist leanings, finally forced the Catholic community to make their serous misgivings known. Where are we heading? Will the government allow a new generation to be poisoned by feelings that run against Brazil's best traditions and the ideals of a healthy society, admittedly without the government's willingness but in any case under its protection?" (from a letter transcribed in Schwartzman, Bomeny & Costa 1984:297–301).

demise. When the university was formally extinguished by a presidential decree in the first weeks of 1939, the project for a new national institution, the Faculdade Nacional de Filosofia, was well under way, and not surprisingly Amoroso Lima had been invited to be its first director.

The Universidade do Brasil was officially established by law on 5 July 1937 along the lines devised by Francisco Campos four years earlier. It would replace the old Universidade do Rio de Janeiro and incorporate its professional schools, some of them dating from the early nineteenth century; and it would be unified by a new Faculdade de Filosofia, Ciências e Letras.

This university was to be the culmination of the large educational empire that Capanema was trying to put together. His speech upon the enactment of the law stressed first that the new institution should "establish the model for higher education in the whole country" and second that it was to be a truly national institution, with students recruited from all over the country on the basis of strict criteria. This was to be an elite university in a completely new university city.[34]

Already in 1935 the minister had formed a committee of fourteen people who were to establish the general plan for the schools, faculties, and other educational institutions and draw up a detailed description of each of those units. Two years later a large amount of work had accumulated and Capanema could announce that the new university, like Minerva from Jupiter's head, was being born complete and ready. A national plan for education was also being prepared and was expected to become law. Capanema hoped that reality would follow from its legal definition: "It is obvious that the establishment of legal norms is not enough. It will be necessary to make the texts come alive, to make them into the reality of the higher education courses in all fields. And this reality will have its standards in the Universidade do Brasil."

Therefore, the Universidade do Brasil was not expected to grow, develop, and find its own way. In fact, it was intended to stifle all attempts at innovation and experimentation in the country, starting, of course, with the Universidade do Distrito Federal. The new university was expected to be "a total, unanimous institution," which for Capanema meant that in its schools and institutes one should find all kinds of teaching prescribed by the laws, so that no higher education institu-

---

34. The following is based on Schwartzman, Bomeny & Costa 1984: chap. 7; quotations are from documents in the Gustavo Capanema archive, Centro de Pesquisa e Documentação em História Contemporânea do Brasil da Fundação Getúlio Vargas, Rio de Janeiro, which have been reproduced or quoted in that book.

tion in Brazil would lack its corresponding model. Higher education would therefore come under strict, centralized control. The Universidade do Brasil was to be as competent as possible, and several innovations, including full-time teaching, were to be introduced. The students would be required to prove their "vocation, dedication, and discipline." There were also plans for libraries, laboratories, museums, and teaching hospitals.

For Capanema the organization of the new university and construction of a university city were almost the same thing, and the same commission was supposed to do both.[35] For the architectural project, Capanema retained the services of Marcello Piacentini, a central figure in Italian fascist architecture who had worked in the planning of Rome's university city. Piacentini's participation was contested by a group of Brazilian architects linked to Le Corbusier (including Lúcio Costa and Oscar Niemeyer, who would become the architects of Brasilia twenty years later), who was also invited to Brazil and made some preliminary projects on his own. In 1938 the models for the university city in Rio de Janeiro were presented publicly in Rome and then shipped to Brazil. But World War II was already too close, and the project was never carried out.[36]

According to Francisco Campos' legislation, a new Faculdade Nacional de Filosofia, Ciências e Letras, was to be the central institution of the new university where scientific research would take place. In 1935 Capanema began work on the project, which would materialize in 1939. It was not to be the first: São Paulo's Faculdade de Filosofia, Ciências e Letras, had been already inaugurated in 1934, and the short-lived Faculdade de Ciências of the Universidade do Distrito Federal began a year later. The invitation to foreign scholars had been a key feature of both the São Paulo and the Rio de Janeiro projects, and Capanema followed the same

---

35. This is how the minister of education assigned the commission's task: "First it should define what the university should be, then it should conceptualize it, then it should project its construction." He knew precisely what he meant by those terms: "Let us assume one determines that the university needs a dentistry school. This is not enough. One should establish also its sections—such-and-such divisions, such a structure. If we establish an institute of criminology, we need to know how many parts, offices, and laboratories it will need" (speech of 22 July 1935, quoted in Schwartzman, Bomeny & Costa 1984:96–97). It was a small step from this to the architectural project.

36. Meanwhile, Lúcio Costa and his colleagues were asked to build the seat of the Ministry of Education in Rio de Janeiro, which was inaugurated in 1945 and has since been heralded as a landmark of modern architecture and evidence of the progressivism and farsightedness of Gustavo Capanema as Brazil's first modern minister of education.

path. One of his advisers was George Dumas, a French intellectual with a long list of contacts in Brazil. Dumas gave Capanema his thoughts about what Brazil needed in terms of higher education and helped to select French professors for the Faculdade Nacional de Filosofia, as he had done years before for the São Paulo school.[37]

In 1936 Capanema wrote to Luigi Fantappié and Gleb Wataghin, Italian professors of mathematics and physics at the Universidade de São Paulo, asking for names. Wataghin replied with a list of Italian professors that included Umberto Nobile from Naples, Giovanni Giorgi from Rome, Beniamino Segre from Bologna, and even E. Schrödinger. None of those would ever come. In 1939 President Getúlio Vargas finally approved the hiring of fifteen foreign professors, and from then on the contacts become official. In June 1939 the Italian ambassador to Brazil confirmed in a letter that the Brazilian government had asked for professors in Italian language and literature, mathematical analysis, rational mechanics, theoretical physics, physical chemistry, superior geometry, and experimental physics and had announced that seven professors had already been chosen and were supposed to arrive in Rio de Janeiro in the following months. Invitations to French professors also went through official channels, with the help of Georges Dumas and under strict ideological control.[38] The French embassy in Brazil helped

---

37. In a 1935 letter to Capanema, Dumas spelled out his views on what Brazil needed in terms of education: "Your youngsters do not lack intelligence . . . , but the good fairy who presides at their birth has also placed in their cribs other gifts besides intelligence: the taste for imagination and dreaming, the abundance of emotional life. These natural dispositions are not to be fought, but it would be very useful if we could limit them to some fields where it is more convenient to have them. Brazil is endowed with admirable poets, certainly because poetry is inherent to your race and all Brazilians are in their own way poets of the soul and nature. But this gift becomes inconvenient when it manifests itself outside the works of poetry and imagination, and the university foundations you are building should moderate and channel these tendencies toward products of reason, from where one should certainly not exclude them completely." He then goes on to say that Brazil is at a critical stage, in which it would be necessary to decide whether it will remain "a country full of charm where all is read and little is produced or whether it will have a place tomorrow among the countries that contribute to the world's intellectual production." The proposed Faculdade de Filosofia should concentrate its efforts on the training of high school teachers and on research, with emphasis on the fields of philosophy, history, and literature (letter of 1 September 1935, transcribed in Schwartzman, Bomeny & Costa 1984:326–29).

38. "For psychology and sociology," wrote the minister to Dumas in 1939, "I need professors used to research and well-oriented studies but related to the church. The Faculdade will be directed by Senhor Alceu Amoroso Lima, a Catholic and friend of

in the negotiations, putting pressure on the minister to speed up decisions, and it did not shy away from reporting the ideological tendencies of French professors.[39]

Following Dumas' recommendations, there was little emphasis on the natural sciences, the request of hard scientists to the Italian government notwithstanding. Capanema had also to accommodate requests from politicians and intellectuals from all sides, and partly because of this Amoroso Lima finally decided to decline the invitation to head the new institution.[40] At this time the Catholic church had already given up the project of bringing the public universities under its control and was beginning to plan for its own Catholic university. Another Catholic conservative, law scholar San Tiago Dantas, took Lima's place and held the post until 1945.

The way the Faculdade Nacional de Filosofia was put together—with its formalism, the lack of intellectual leadership, its ideological prejudices, the reliance and dependency of bureaucratic procedures at each step—extracted its toll. In spite of a few outstanding names, it would never become a significant center for scientific research—or even a center of conservative thinking in the social sciences and humanities, for which it had been conceived.

Jacques Maritain. This is why I would not be pleased with names known for tendencies that are opposed to or divergent from those of the church" (letter of 17 July 1939, quoted in Schwartzman, Bomeny & Costa 1984:216).

39. Of Professor Poirier, who was supposed to teach philosophy, the ambassador affirms, "Il a déjà indiqué que son orientation doctrinale répondait entièrement aux vues du directeur de la nouvelle faculté" ("He has already indicated that his position corresponds totally to the views of the director of the new faculty"). About Professor Ombredonne, for psychology, the embassy confirms that he "présente toutes les garanties souhaitables au point de vue des tendances" ("has all the desirable qualities with regard to inclination"); the person for sociology is Jacques Lambert, who had been in Brazil before and who is said to belong to the same generation of Catholic professors as Deffontaines and Garric (quoted in Schwartzman, Bomeny & Costa 1984:216).

40. In April 1939 Amoroso Lima was still willing to do it, but only if the new school did not bring in the almost 100 faculty and 500 students left idle by the closing of the Universidade do Distrito Federal. Three months later he decided the new school would never be the way he wanted and wrote an angry letter to Capanema protesting the designation of professors "by others, not us," and more specifically the designation of an "unknown Italian" to teach physics in the place of Joaquim Costa Ribeiro. In January 1941 Lima formalized his decision, arguing that it would be impossible for him to dismiss many professors who, openly or not, are establishing a climate of "philosophical and ideological confusionism" (quoted in Schwartzman, Bomeny & Costa 1984:218).

A New Elite for a New Nation: The Universidade de São Paulo

The establishment of the Universidade de São Paulo in 1934, by contrast, is the most important event in Brazil's scientific and educational history. To understand how it was created, how it could develop so differently from the institution in the country's capital, and what role it was to play in the future, we should look at it in the light of São Paulo's defeat in the 1932 uprising against the Vargas regime. A central figure in the project was Júlio de Mesquita Filho, owner of *O Estado de São Paulo* (a traditional newspaper dating back to the nineteenth century) and a relative by marriage to Armando de Sales Oliveira, the state's leading politician in 1932, who returned from exile two years later to become the state's federal intervenor under the provisional government of Getúlio Vargas. Another important name was Roberto Simonsen, business entrepreneur, author of the first modern economic history of Brazil, and leader of the state's Federação das Indústrias. Thus, the supporters of the idea of a state university were among the most powerful members of the local agricultural and industrial elite. This was in sharp contrast to Capanema's projects in Rio de Janeiro, which were marginal for the Vargas regime and never got much support or generated much interest outside the circles of educators and the church.[41]

The concerns of São Paulo's elites with technical knowledge and higher education did not begin in 1932. There were already a few fairly successful institutions in their state—the Instituto Butantã, the Instituto Biológico, the schools of engineering and medicine—but they wanted to make it more directly relevant to the management of the economy and for the nation as a whole. The concern with scientific management had already prompted the state's Associação Comercial to create the Instituto de Organização Racional do Trabalho (Institute for Rational Organiza-

---

41. "With Armando de Sales in power and Júlio de Mesquita Filho as the director of *O Estado de São Paulo*, we believed the moment had come to create the Universidade de São Paulo and the Faculdade de Filosofia, Ciências e Letras. Júlio de Mesquita and I have been fighting for this since 1923. In that year, and in 1926, I wrote several articles on the subject for *O Estado* and in 1925 I initiated a large survey, which took several months, about public instruction in São Paulo, in which we dealt with and discussed the problems of higher education in our state.... Given the crossroads in which we found the state's educational system, we believed we needed radical solutions, from the top down, including the creation of a university and its Faculdade de Filosofia, Ciências e Letras" (F. de Azevedo 1971:119–20).

tion of Labor) through the initiative of Roberto Simonsen and with Armando de Sales Oliveira as its president.[42]

The idea was expanded by the same people in 1933 with the creation of the Escola Livre de Sociologia (Free School of Sociology). The Escola was preceded by a proclamation published in all São Paulo newspapers and signed by the directors of all higher education institutions in São Paulo and a large list of well-known personalities.[43] It stated the intention of making the new institution "a center for political and social culture capable of fostering interest for the common good to establish the connections between people and their environment, to stimulate research on the living conditions and the vital problems of our populations, and to shape personalities able to participate efficiently and with self-awareness in the leadership of our social life." These personalities were to provide for what the proclamation called "the lack of an organized and large elite, trained through scientific methods, aware of the institutions and achievements of the civilized world, and able to understand first and act later on our social environment." The absence of this elite was directly related to the frustrations of the 1932 rebellion.

The Escola Paulista de Medicina was also created in 1933 and was expected to bring radical innovations to Brazil's higher education traditions. The new institution was to provide high-quality teaching, to promote biomedical research, and to open new opportunities for medical education in the state, which so far had been limited to the small number of students recruited each year by the Faculdade de Medicina de São Paulo. One important novelty was that the Escola Paulista was supposed to get both private and state support, which it did until its federalization in the postwar years.[44] The Escola Paulista de Medicina was a success in terms of its original intentions. It still ranks among the country's most prestigious medical schools. The Escola de Sociologia e Política, meanwhile, dwindled after some years of intense activities,[45] and it was never able to establish an academic tradition similar to the one developed at the Faculdade de Filosofia, Ciências e Letras.

These two initiatives were followed by the creation of the university, which was meant to be the state's best response to its military defeat in 1932.

42. It was preceded by the Instituto de Organização Científica do Trabalho (Scientific Organization for Labor), directed by a Swiss specialist in industrial psychology, Leon Walter, a first experience that did not last long.
43. Berlink and Ferrari 1958.
44. Albernaz 1968; Vale 1977; Pena 1977.
45. L. L. Oliveira 1986.

Defeated by the strength of arms, we knew perfectly well that only through science and continuous effort could we recuperate the hegemony we had enjoyed for several decades in the federation. Paulistas to the bones, we had inherited from our ancestors, the bandeirantes, the taste for ambitious projects and the patience needed for large undertakings. What larger monument than a university could we build for those who had accepted the supreme sacrifice to defend us against the vandalism that had just desecrated the work of our elders, from the bandeiras to independence, from the Regency to the Republic? . . . We came out of the 1932 revolution with the feeling that destiny had placed São Paulo in the same spot as Germany after Jena, Japan after its bombardment by the American navy, or France after Sedan. The history of those countries pointed to the remedies to our evils. We had experienced the terrible adventures caused, on one hand, by the ignorance and incompetence of those who before 1930 had decided on the destiny of our state and our nation, and on the other hand, by the emptiness and pretentiousness of the [1930] October revolution. Four years of close contacts with leading figures of both factions convinced us that Brazil's problem was above all a question of culture. Hence the foundation of our university and later the Faculdade de Filosofia, Ciências e Letras.[46]

The new university would be public, lay, and free from religious influence; it was to be an integrated institution, not simply a gathering of isolated schools. Its core was to be the Faculdade de Filosofia, Ciências e Letras, made up of professors invited from abroad. There research would be done by a full-time staff that would work on higher forms of science and leave the practical chores to the professional schools. The university was to be autonomous from both the administrative and academic points of view, and it was destined to produce a new elite that would take up the country's leadership, overcoming its backwardness and bringing São Paulo back to its deserved position as the country's leading state.

Beyond these general outlines, it was necessary to choose an organizational model to follow. The Francisco Campos legislation had already established the notion of a central Faculdade de Filosofia for scientific work and teacher education, drawn probably from the Italian experience, and this was maintained. The Paulistas like to think that this was their invention. One of the organizers, Paulo Duarte, recalls:

---

46. Mesquita Filho 1969:164, 199. The text is from 1937.

> We selected two paradigms, so to speak, for the university. First, both Julinho [Mesquita Filho] and I have a French education, but we did not want to limit ourselves to our French education. We selected the Sorbonne, of course, as a model for a scientifically organized university. We also selected the English model, through Cambridge. We sent for as much information as we could about these universities. But the French organization was better than the English, so we can say our organization followed 80 percent of the French model. . . . Earlier the faculties of philosophy, sciences, and letters were the *cellula mater* of the French university. Later they were divided into the schools of philosophy and letters on the one hand and the schools of science on the other. We did not have the means to make two schools from the beginning, so we decided to do as the French did in the past. All other schools revolved around this one. I do not recall the English structure well, but I do recall that in England science was completely separate from philosophy. For the more advanced sciences, such as biology, they had independent institutes. France also had institutes outside the universities; they were only associated with them.[47]

Besides the sketchy knowledge about the true characteristics of the academic systems supposedly taken as models, Duarte's statements suggest a prevailing concern with organizational forms, not with the academic and scientific quality and achievements of the institutions whose features they were about to borrow. In part this was because both Duarte and Mesquita were *hommes de lettres* in the French tradition—rather than scientists—and above all political activists. Duarte defined himself as "democratic socialist," and in that sense was politically marginal, while Mesquita was a liberal in the classic tradition and very much a member of the Paulista establishment.

These statements reveal also the restrictions under which the project was carried on. The general outline of the Francisco Campos legislation had to be followed, and that included a school of sciences and education. Also, the traditional higher education and research institutions in the state were to be incorporated in the project, and they could not be easily tampered with. The assumption was that these institutions would resist any form of integration that went beyond simple juxtaposition and autonomous coexistence or that attempted to go beyond the simple benefits of bureaucratic and material rationalization.

The bill establishing the university was signed by the state governor on

---

47. Duarte interview.

25 January 1934.⁴⁸ In contrast with the Francisco Campos legislation, this bill was brief and written in clear and direct language. It had only fifty-four articles, as opposed to the 328 items in the Francisco Campos legislation for the Universidade do Rio de Janeiro. The first goal of the new university was "to promote the progress of science through research"; the second goal was to transmit knowledge; the third was to form specialists and professionals; and the fourth goal was to promote the diffusion and popularization of the sciences, arts, and letters through short courses, conferences, radio broadcasts, scientific films, and so on.⁴⁹

The Paulistas talked about their university, but most of their efforts went into creating its new Faculdade de Filosofia, Ciências e Letras. Given the background of the founders, it was natural to expect that they would emphasize the social sciences and the humanities. Only afterward, when Teodoro Ramos, Rocha Lima, and other scientists were invited to join the Faculdade de Filosofia's organizing commission, did the natural sciences receive more attention.⁵⁰

The project was to recruit the full faculty in Europe, with a special place for France. With European fascism on the rise, France was per-

---

48. Fernando de Azevedo, who helped draft the bill establishing the university, recalls the events: "Júlio de Mesquita Filho . . . , in his and Armando de Sales' behalf, asked me to draft the bill that would create the Universidade de São Paulo. The final version was completed in less than four days, including the introduction and the justification. It was December 1933, and I asked Mesquita to plead with Armando de Sales not to sign the bill immediately. . . . Because the new university would include the already existing professional schools in law, engineering, medicine, and agriculture, it would not be prudent to sign the decree without hearing from them first. To break down the resistance, a fourteen-member commission was established—two from each school, two from the Instituto Agronômico de Campinas, two from the Instituto Biológico de São Paulo, and two from the Faculdade de Educação. . . . The commission worked for fifteen days and approved my project with some minor restrictions" (F. de Azevedo 1971:120–21). Paulo Duarte, who participated in the group, discovered at the last minute that his name had been dropped from the commission's list, for reasons that remained unclear (Duarte 1976:71–73).

49. State Decree no. 6283, 25 January 1934.

50. Paulo Duarte describes the establishment of this commission as a negotiation between himself and Júlio de Mesquita. "The commission was formed by Henrique de Rocha Lima and Fernando de Azevedo, my candidates, who were accepted by Julinho with reservations: the first because he was too German, the second because he was a former seminarian, which stripped him of all serenity. Teodoro Ramos had been an instrument of São Paulo's enemies but was extremely intelligent and one of the few among us who could teach higher mathematics in a university. There were doubts about Raul Briquet, whom Julinho believed did not know what a university really meant. The same thing happened with Agesilau Bittencourt—suggested by Rocha Lima, who was not supposed to have enough general culture. There was no discussion about the others: Vicente Rao, Fonseca Teles, André Dreyfus, and Almeida Júnior" (Duarte 1976:68).

ceived as a liberal alternative and in tune with Brazil's tradition of French influence in the humanities.

> We wanted to make use of the best not just from one advanced country but from all advanced countries. Thus, Italy was to provide professors of mathematics, geology, physics, paleontology, and statistics; Germany would provide those in zoology, chemistry, and botany; England could help in another branch of natural history and perhaps psychology; and for France would be reserved the chairs of pure thought: sociology, history, philosophy, ethnology, geography, and perhaps physics. It was not always possible to meet this plan.[51]

> The presence of a large Italian colony in São Paulo and the Italian government's eagerness to help posed special problems. We could not let the chairs of the Faculdade de Filosofia fall into the hands of followers of the Italian creed, especially those chairs more likely to influence the moral education of our youth. The difficulty was compounded by the large number of children from the peninsula who lived in São Paulo, most of whom did not hide their tendencies to accept the orientation of fascist Rome. We had to be extremely careful because of the growing and impertinent pressures the Italian government and the Italian colony were placing on the Paulista government. They wanted to force the arrival of a large number of Italian professors to make up the new faculty. We solved the problem by offering the Italians some chairs in pure science (mathematical analysis, geometry, statistics, geology, mineralogy) and in Italian language and literature. We kept for France, leader of liberal democracy, those more directly related to the spiritual education of the future students: philosophy, sociology, political economy, politics, human geography, classical literature, and French language and literature. The rest—chemistry and natural history—were to be filled in by Germans being expelled from their country by Hitlerism. Thus, we could protect the liberal meaning of the Brazilian evolution.... Future elites would not be victimized by the teaching, through the chairs, of those exotic theories that went against the nature and inborn tendencies of our people.[52]

51. Duarte 1976:70.
52. Mesquita Filho 1969:192. It is not true that a significant part of the new school's faculty was made up of political refugees and there were almost no Jews.

The Francisco Campos legislation intended the faculties of philosophy to be geared toward the education of high school teachers, putting scientific research in the distant future. In the Universidade de São Paulo, by contrast, the intention was to establish a rigid divide between the Faculdade de Filosofia and the Instituto de Educação, the teachers' college. The Faculdade de Filosofia would provide the students with substantive knowledge, and those interested in teaching would have to go to the Instituto de Educação for pedagogical studies and licensing. The teacher would be a by-product of the new institution.[53]

The idea was thus to establish academic activities in two tiers. The lower tier was to be geared to applied work and professional education and would include the old professional schools of medicine and engineering; the higher tier would cover all scientific specialties needed for the former plus those fields deemed essential parts of human knowledge. The higher level was supposed to nourish those in the lower one and gradually raise their levels of proficiency. No foreign professors were hired for the old schools, which changed very little in the years to come. The beneficial influence that was supposed to reach them from the new Faculdade de Filosofia turned into a source of permanent conflict and resistance to change, which sometimes even threatened the survival of the Faculdade de Filosofia and the whole university project.[54] A source of conflict was the idea that basic disciplines, such as mathematics, chemistry, and biology, were to be organized in central chairs linked to the Faculdade de Filosofia, which would then provide the professional schools with the courses they needed on these subjects. This integration would require bringing together all institutions on the same campus, a

---

53. We wanted "an institute where nothing else [but science] would be done, where the true vocations could find an endless area to expand their inborn tendencies, where the rule would be science for science's sake and the spirit of research could dominate all minds. In a word, we would fill the immense gap in the nation's culture by giving academic studies their rightful leading place in the intellectual hierarchy or a university organism" (Mesquita Filho 1969:189).

54. "We had lengthy discussions about the convenience of bringing into the new university such institutions as the Escola Politécnica, Faculdade de Direito; the pharmacy, dentistry, and veterinary schools; and the Luiz de Queiroz [agriculture] from Piracicaba—all of which are outdated, loaded with superiority and inferiority complexes, and without any understanding of what a university really is. . . . Julinho pointed out the danger that the old schools would contaminate the new ones with their incurable vices. They were all resistant or even allergic to research, their pedagogy was obsolete, and the simpleminded vanity of their self-taught professors would resist all attempts to change them. Other members of the commission, primarily Rocha Lima and Dreyfus, reminded us that the opposite also could be true and that this would be the case if we kept the university under strict surveillance in its first ten years" (Duarte 1976:70).

long-term project that was not considered at the time. The authorities of the Faculdade de Direito stated, from the onset, that they would not accept a transfer to a university city if such a city was ever built. The Faculdade de Medicina refused to allow the construction of a new floor in their building to house some sections of the Faculdade de Filosofia. The Politécnica refused to have "philosophers" teaching their basic disciplines. When Luigi Fantappié agreed to go to the Politécnica to teach, he was charged with not being competent enough. The idea of unified departments and institutes was set aside and remained dormant for several decades.[55]

The decision that the entire faculty of the new school should come from abroad was a radical one. At first, Teodoro Ramos was considered for the chair of mathematical analysis; André Dreyfus for general biology; and Fernando de Azevedo was considered for sociology, according to Azevedo himself. Ramos already held the chair of mathematical analysis at the Politécnica, and Dreyfus was considered the best and most broad-minded biologist in Brazil. They all decided, however, that they needed at least one or two years of study abroad before taking up such a task.

There is no single register of the persons invited, those who actually came, or the duration of their stay. The first group included, from France, Paul Arbusse Bastide (sociology), Émile Coornaert (history of civilization), Robert Garric (French literature), Pierre Deffontaines (geography), Etienne Borne (philosophy and psychology), and Michel Berveiller (Greco-Roman literature); from Italy, Francesco Picollo (Latin), Luigi Fantappié (mathematical analysis, integral and differential calculus), Ettore Onorato (mineralogy), and Gleb Wataghin (theoretical physics); from Germany, Ernest Breslau (zoology), Heinrich Rheinboldt (chemistry), Felix Rawitscher (botany); and from Portugal, Francisco Rebelo Gonçalves (Portuguese literature).

Besides those, the Faculdade's first yearbook, for 1934-35, lists Jean Mougé, Pierre Monbeig, Fernand Braudel, Claude Lévi-Strauss, Edgar Otto Gothsch, and Pierre Hourcade, all from France, and also the first Brazilian names: Teodoro Ramos (who was also the school's first director), Luís Cintra do Prado, Antônio Soares Romeu, André Dreyfus, Paulo Sawaya, Afonso d'Escragnolle Taunay, and Plínio Airosa. Those listed as technical assistants were Omar Catunda, Ernesto Luís de Oli-

---

55. Mesquita Filho 1969:172-73. A large campus was built for the university several decades later, but neither the medical school nor the law school ever joined it, and as late as 1988 the faculty of the medical school threatened, in a dispute about a change in the university statute, to quit the university and recover the autonomy it lost in 1934.

veira, Fernando Jorge Larrabure, Heinrich Hauptmann, Herbert Stettiner, Reinaldo Saldanha da Gama, Maurício Rocha e Silva, and Gertrud Siegel. A second wave brought Ernst Marcus, Paul Vanorden Shaw, François Perroux, Luigi Galvani, Giacomo Albanese, Giuseppe Ungaretti, Georges Readers, and Ottorino de Fiori Cropani. Marcus was to replace Breslau, who died suddenly. Most professors came for a short period and returned after the first year. They were often replaced by others from the same country, such as Jean Gagé, Pierre Fromont, Roger Bastide, Alfred Bonzon, Karl Arens, and Atílio Venturi.

In the first years the sections of physics and mathematics were at the Politécnica, while the others were placed at the Faculdade de Medicina. Enrollments for the first courses were open in early 1935, but the new Faculdade meant little to the young Paulistas finishing high school and even less for the children of the city's leading families. They wanted a prestigious profession, but the Faculdade de Filosofia offered nothing of the kind. As a consequence, enrollment was minimal, and the solution was to look for students who were applying to the Instituto de Educação (which had Fernando de Azevedo as its director) and who were granted provisional enrollment in the new Faculdade. Classes started on 11 March 1935 with forty-six students in philosophy, twenty-nine in mathematics, ten in physics, twenty-nine in chemistry, fifteen in the natural sciences, sixteen in geography and history, eighteen in the social and political sciences, five in classical literature and Portuguese, and nine in other foreign languages.[56]

Although the Paulista elite did not send their sons to the new Faculdade, they could attend the lectures, and the conferences of the best-known visitors attracted a large public, even sometimes intimidating the small number of regular students. It was not usual to see Armando de Sales de Oliveira, Júlio de Mesquita, or their relatives in the audience.

Claude Lévi-Strauss provided a rare outside view of these events. He had been a student of Georges Dumas and came to Brazil with his help (and not, as Paulo Duarte reconstructed so many years later, against his teacher's will). Brazil was Lévi-Strauss' introduction to ethnography and the tropics, and he preferred the authenticity of the Indians to the Paulistas' attempts to mimic European science and civilization. In his recollections he talks about American cities that go from barbarism to decadence without getting civilized and about the Brazilian ambassador in Paris who wanted to appear civilized and denied that there were any Indians left in Brazil. He saw the new university as a response to the need of the Brazilian oligarchy to form "a public opinion inspired by

56. F. de Azevedo 1958:222; E. de S. Campos 1954:427.

civilian and lay values, to compensate the traditional influence of the church, the army, and of personal power." To do so, it was necessary to provide culture to a larger public. For Lévi-Strauss, all the excitement around the foreign professors was viewed with suspicion: there were people trying to get jobs or fend off future competition, and rites of status and prestige in exhibitions of familiarity with the latest Parisian intellectual fashions and their representatives. Nobody was really concerned about the intrinsic value of ideas and concepts being tossed around. With diplomas granted by the university, these newcomers would have access to administrative jobs, thus forming a new elite to replace the feudal arrangements that existed until then. This was, for Lévi-Strauss, the most obvious product of the French cultural mission to Brazil, something that Georges Dumas, fascinated with his powerful friends in the Brazilian oligarchy, never realized.[57] Without really understanding or caring about all this, Lévi-Strauss joined as soon as possible what was for him the more authentic reality of the Caduveo, Bororo, and Nambikwara tribes in the Brazilian jungle. Only the "virile resolution of a small group of gifted children"[58] could explain how his students—in large part women—could become a significant community of social scientists fifteen or twenty years later.

The Universidade de São Paulo was in many ways a frustrated project. The expected integration among the professional schools did not happen; enrollment at the Faculdade de Filosofia was always difficult, and it remained in many ways a teachers' college; its students tended to be women, people from small towns outside the state capital, or children of recent immigrants. It was impossible, under these circumstances, to have the new institution playing the roles of elite formation and leadership, which Júlio de Mesquita expected. If there was a hierarchy of prestige and recognition among the different institutions that were brought together at the Universidade de São Paulo, the Faculdade de Filosofia was not at the top. With the Estado Novo and the political ostracism of Armando de Sales Oliveira and his group after 1937, the Universidade de São Paulo came under pressure of all kinds, emanating both from the new state authorities and from the Ministry of Education in Rio de Janeiro.

In spite of all that, the new university—and more precisely its Faculdade de Filosofia—became the most important scientific institution ever established in Brazil after the Instituto Oswaldo Cruz. This is partially explained by the economic conditions of the state, which could

57. Lévi-Strauss 1955:114, 118.
58. Lévi-Strauss 1955:117.

provide it with more resources than any other similar institution in the country could ever get. But it would be a mistake to try to explain everything by the power of money. More important was that even on a small scale the Faculdade de Filosofia did provide a space for science, to be carried on by a small group of foreign visitors and their Brazilian disciples; and the ambitious goals of Júlio de Mesquita, even if largely frustrated, helped to preserve the respectability and recognition of the new institution throughout the years. Even the recruitment pattern of the Faculdade de Filosofia turned out to be a blessing. Scientists are rarely recruited at the top of the social hierarchy, where money and power are so much easier to get. For the women, the children of immigrants, and the country people who came to the new institution, often hoping for little more than a teaching job, science became suddenly a new window to the world. Many of them grasped it with enthusiasm.

An obvious weakness of the São Paulo project was its isolation from the English-speaking intellectual and scientific environment. Contacts with England and the United States became more intense during the war, and they dominated afterward.

The question of the university model adopted by the Universidade de São Paulo is an interesting one. France was the prevailing influence, and in the past Brazilians had tried to copy some of the *grandes écoles*, such as the Politechnique and the École de Mines. But there had been nothing similar to a faculty of philosophy in France since the French Revolution. Cambridge, although mentioned by Paulo Duarte, was never really considered as an alternative. Italy, which is not mentioned by the Paulistas, was the source of much of Francisco Campos' and Gustavo Capanema's educational ideas and had a similar organizational model.

What prevailed at the Universidade de São Paulo was not so much what its founders had put down on paper but what the foreign visitors drew from their experience. The Germans continued with their research in chemistry or biology as they had traditionally done in their country, but for the educators the Faculdade de Filosofia remained always a kind of *école normale*. Each institution brought into the university also kept its own organization and traditions. The Faculdade de Medicina was close to the North American pattern, thanks to the support it received from the Rockefeller Foundation; the Escola de Direito never changed its professorial style; the Escola Politécnica remained an institution geared to technology and resisted the incorporation of modern physics; and so on. This mixture of different academic models, traditions, and experiences within the same institution became in time one of the strong points of the Universidade de São Paulo, where centralization and bureaucratic domination was never fully to prevail.

6

THE ROOTS OF
SCIENTIFIC TRADITIONS

Most of contemporary science in Brazil has its roots in the scientific traditions and institutions established and strengthened in the transition years of the 1930s. Some of them, like those in the biological sciences, grew out of applied research institutes of the previous years; others, like those in modern physics, started with the Universidade de São Paulo. As a rule, only those fields that could be organized academically survived the 1930s to come alive in the 1950s, 1960s, and later years. Those that did not—like the earth sciences or technological research—had to wait for new beginnings several decades later. In this chapter I reconstruct in some detail the path followed by some of these traditions and show how they helped to consolidate the foundations on which Brazilian science was supposed to grow.

From Agronomy to Genetics

Agricultural research began in Brazil (if we disregard the old botanical gardens) with the creation of the Estação Agronômica de Campinas, which was to study tropical plants under the direction of the Austrian chemist F. W. Dafert.[1] Campinas was in those years at the heart of coffee country, but neither the inhabitants of that region nor Brazil's minister of agriculture, who created the Estação Agronômica, were very appreciative of Dafert's studies on coffee fertilization, and in 1890 he was dismissed, being reinstated some time later thanks to the intervention of another foreign scientist, Orville Derby, head of São Paulo's Comissão Geográfica e Geológica.[2] The station's name changed to Instituto Agronômico de Campinas, and it came under state jurisdiction in 1892; in 1907 Dafert was finally removed after a failed attempt to transform the Instituto Agronômico into a purely practical institution concerned with solving short-term agricultural problems and operating as a lucrative business.[3]

The Instituto Agronômico thus entered the 1920s removed from the first goals laid out by Dafert, as illustrated by its reduced staff: one head agronomist, two gardeners, and some laborers. In 1927 Teodureto de Camargo undertook a reform that was basic in attempting to return to its original philosophy: the study of agricultural problems was to be carried out first in laboratories and experimental fields and later at the institute's various substations located around the state; only at a third moment would the results be released. By the 1930s the Instituto Agronômico

---

1. Dafert was born in Vienna and had a doctorate from Giessen, Germany. In 1898, after returning to Europe, he was appointed director of Vienna's Experimental Chemical-Agricultural Station. For a full study of Dafert's presence in Brazil, see Dean 1989.

2. "The beginning of experimental work at the Estação Agronômica not at all surprisingly attracted the keen interest of both large plantation owners and those who were simply curious about what the specialists were doing there. But a few were perplexed by what they saw. The work being carried out seemed too theoretical; perhaps it could somehow be of use to national agriculture, but only in the future. As much as the director of the station tried to explain that such research was necessary for him to determine what direction to take in future experimental work, he could not manage to convince them. Rumor spread that the director was conducting experiments purely for his own scientific motives, without concern for immediate practical application" (F. Campos 1954:496).

3. "The observations and experiences that had so far been accumulated were inadequate to justify dissemination among plantation owners, as they still needed to be submitted to tests in various regions of the state whose climate and soil differed from that of Campinas. Only then, if results so indicated, could the relevant advice be given to interested coffee growers. As this basic principle was ignored, São Paulo plantation owners suffered many serious losses" (F. Campos 1954:497, 498).

was operating again at full steam, mainly because its specialists had all been designated full-time workers in 1929. Its studies had repercussions in the academic world and were important in improving São Paulo's agriculture. With coffee in a period of crisis caused by overproduction and international recession, the institute provided local farmers with cotton seeds and other crops.

The 1930s were also an important decade for the Escola Superior de Agricultura Luiz de Queiroz, created in 1901 as the Escola Agrícola de Piracicaba under São Paulo's state secretary of agriculture.[4] The goal was to provide agricultural education at all levels, from primary schools to graduate studies, in the same educational and research establishment, thus providing continuity and coherence in the training of specialists.[5] The quality of the work done by the Escola Superior de Agricultura Luiz de Queiroz in basic and applied science within the field of agriculture is illustrated by its pioneer decision to teach genetics in 1918 as part of the agriculture class led by Carlos Teixeira Mendes and as part of the zootechnics class led by Otávio Domingues. It was the first time that this new subject was taught in Brazil.

Only in 1928, however, was genetics used systematically by the Instituto Agronômico de Campinas for the improvement of products like coffee, corn, and tobacco or for the adaptation of others, like wheat and barley, to the Brazilian environment. In 1932 C. A. Krug was sent by the Instituto Agronômico to Cornell University to specialize in genetics, cytogenetics, and the improvement of plants. He returned to Brazil by the end of 1932 and organized a research group to work in the improvement of coffee and

4. The Escola owes its existence mainly to Luiz Vicente Sousa Queiroz, a wealthy patron of Piracicaba who had already provided the city with such benefits as electric energy. In 1892 Luís de Queirós took advantage of a state law that created the Escola Superior de Agricultura with ten subordinate experimental stations. He donated to the state his São João da Montanha ranch in Piracicaba, to be used as the site of this Escola. Initially backed by two important state politicians, Jorge Tibiriçá Piratininga and Bernardino de Campos, construction of the Escola nonetheless came to a standstill when the former left his post as state secretary. The Escola was inaugurated only in 1901, and then not as the Escola de Agronomia but as the Escola Agrícola Prática. When Jorge Tibiriçá won the governor's seat, it was finally possible for the Escola to be set up as originally conceived.

5. While the Escola sponsored visits by such outstanding foreign professors as Nicolas Athanasov, Arsène Putmans, and others, travel grants were created under Governor Rodrigues Alves, thus providing further training for such students as Carlos Teixeira Mendes, Trajano Sampaio, and José de Melo Morais. The Ministry of Agriculture was prompted to offer similar grants to graduates of agronomy schools, allowing such students as José Vizioli and Salvador de Toledo Pizza Júnior to broaden their knowledge of various fields through studies in Europe or the United States.

corn. A chair of genetics was introduced in the next year with the purpose of training specialists in the technology of plant improvement.

The Escola Superior de Agricultura Luiz de Queiroz was to follow a different approach by inviting Friedrich Gustav Brieger to organize its department of genetics. Born in 1900, Brieger received a doctorate in botany from the University of Breslau in 1921, and in the next few years he worked at the universities of Munich, Berlin, and Vienna. In 1924 he received a fellowship from the Rockefeller Foundation and worked for two years at Harvard with Edward M. East, whom he considered his strongest influence. From Harvard he became a researcher at the Kaiser-Wilhelm-Institut, where he worked with Karl E. Correns, known for the rediscovery of Mendel's laws. In 1933 he left Germany for England, where he worked at the John Innes Institute until being invited to organize the genetics department of the Escola Luís de Queirós. Encouraged by opinions of other Europeans who were also coming to São Paulo at the time, he decided to embark on his "tropical adventure," as he himself called it.[6]

Brieger's achievements should be credited to his academic qualifications and to his ability to work with people who, although not scientists themselves, were engaged in the creation of a new academic mentality in São Paulo. He stressed the role of José Melo Morais, director of the school upon his arrival.[7] There was also André Dreyfus, more an intellectual and self-taught man than a researcher, who would play a central role

---

6. Brieger recalled that when he arrived he "encountered a very interesting situation ... in Piracicaba. José de Melo Morais, director of many years, was an exceptional fellow. He was a chemist and had studied in Germany, and although he was not a researcher his intuition was excellent. He had realized that Brazil's old teaching system, based on the book and ignoring research, would no longer do. Taking up the banner of the University of São Paulo, he wanted to make Escola Luís de Queirós a full-time course and introduce research, in order to turn a teaching school into a university institution. I found this all extremely favorable, because I just can't imagine any university teaching without research" (Brieger interview). Brieger later stated: "Something else that was very favorable at that time was that André Dreyfus in São Paulo had taken a post as head of the Departamento de Biologia Geral and was also interested in introducing genetics. And Carlos A. Krug in Campinas, head of the genetics section that he himself created, began introducing genetic improvement methods—I mean, improvement on scientific bases. We three established a wonderful friendship and a great work atmosphere, so that we ourselves criticized each other and defended each other from the rest. We three had the idea not only to use fundamental and applied methods but also to train disciples."

7. "He was far from a researcher, but he could smell, and he had perceived that the old Brazilian way of teaching from the book, without research, did not work. He joined the project of the new university in the quest for full-time work and research and really moved in the direction of changing a technical school into an academic institution" (Brieger interview).

in the introduction of modern genetics at the Universidade de São Paulo.[8]

Brieger, Krug, and Dreyfus made a scientific community in miniature and set themselves to the task of not only doing research but, mostly, training disciples and creating a scientific tradition. While Krug worked mostly with applied genetics with established methodologies, Brieger was most interested in coming up with new approaches. His first work in Brazil was with corn and lettuce. In his studies of corn he was the first to utilize the genetic analysis of populations, instead of hybridization, as a technique for species improvement. To do that he had to make use of sophisticated mathematical models, and his insistence on the superiority of this approach to the traditional one led to a conflict with Krug and the end of their collaboration. His work with lettuces led to the refutation of old teachings that were taken as scientific truths.[9]

In 1938 Dreyfus began work full-time at the department of general biology of the Faculdade de Filosofia. Both the old Instituto Agronômico and the Escola Luiz de Queiroz had been incorporated into the new university, and Krug and Brieger remained where they were outside the city of São Paulo. At the Faculdade de Filosofia, Dreyfus and three assistants—Martha Brener, Crodowaldo Pavan, and Rosina de Barros—strove to improve the quality of their work. A real change came, however, only a few years later, with the arrival of Theodosius Dobzhansky.

Dobzhansky arrived in Brazil in 1943 with the support of the Rockefeller Foundation. In 1936 he had published a book that was widely regarded as one of the most important contributions to genetics since Darwin. He had applied for a grant to go to Central America and was persuaded by Harry Miller, an adviser to the Rockefeller Foundation well acquainted with Brazil, to come to São Paulo. He is remembered as extremely energetic, and he changed the sedated rhythm of the Brazilians with his constant requests for field trips, grants, and equipment. Dreyfus not only did not compete with him, but became Dobzhansky's main supporter and defender.[10] In São Paulo, Dobzhansky developed a

---

8. "He would do very little research himself, but was able to absorb knowledge and transmit it to other people; he therefore had the quality needed to create a school, and he did" (Brieger interview).

9. "The scientific belief was that green vegetables are from a temperate climate and could not be planted in the tropics except at high altitudes. Seeds particularly had to be produced at high altitudes. I wanted to work in the improvement of plants, and since we did not have enough altitude in Piracicaba I decided the plants would have to adapt to me. I knew by experience that many of those scientific theories were just a consequence of lack of observation. Nobody knew much about the tropics in those years" (Brieger interview).

10. Pavan interview.

research line on the population genetics of the drosophila that received quick international recognition. Later, several of his students and assistants went on to complete their education in the United States. These formed a network of genetics specialists—working not only in São Paulo but in Porto Alegre, Brasília, and Paraná—in the fields of genetics of human populations, cytogenetics, and medical genetics. Students of Brieger stayed closer to agricultural research and developed studies in the genetics of bees and fungi.

Out of Manguinhos:
The New Institutes of Biological Research

A parallel development was the 1927 creation of the Instituto Biológico de Defesa Agrícola e Animal de São Paulo (Biological Institute for the Protection of Agriculture and Animal Husbandry) to replace an old commission for the study and eradication of coffee borer, a plague of which was threatening the state's main agricultural product. The new institute was a direct outgrowth of the research traditions of the Instituto Manguinhos in Rio de Janeiro. Its first director, Artur Neiva, worked with Oswaldo Cruz in the first sanitary campaigns and headed the fieldwork in the fight against malaria in different regions.[11]

The new institute started with mandates to do applied and basic research, to promote protective measures for the state's agricultural products, to teach their proper use, and to produce serum and inoculations against animal diseases. The staff included agronomists, veterinarians, medical doctors, biologists, and chemists who were organized in two divisions, one dealing with plants, headed by Adalberto de Queirós Teles, and the other dealing with animals, headed by Henrique de Rocha Lima.[12] The institute recruited, among others, the first three graduates of Manguinhos' "curso de aplicação" in the year of its foundation: Otto

---

11. According to José Reis, Neiva was equally comfortable in his laboratory and in the field, and he became one of the most important entomologists of his generation. As director of hygiene in the state of São Paulo, he drafted the first sanitary code in Brazil's history. He headed the old commission and was responsible for transforming it into a permanent institution (Reis 1976b and 1976d).

12. The first division was subdivided into sections of botany and agronomy, chemistry, entomology and parasitology, and phytopathology. It included most of those who had worked on the old commission plus newly recruited botanists, agronomists, entomologists, and chemists. The animal division, headed by Genésio Pacheco, was divided into the sections of physiology, bacteriology, anatomy, pathology, and entomology and parasitology.

Bier, José Reis, and Adolfo Martins Penha. Its eight sections were supposed to be in constant contact with each other, sharing a good library and technical services. From the onset the idea was that applied and basic research should coexist in harmony, as it did in Manguinhos in its best years.[13]

Otto Bier characterized the institute at its beginnings as marked by "a critical mass of people interested in the same goal, namely to carry on with serious scientific work silently without any concern with self-promotion and with a strong sense of continuity."[14] The institutional culture assumed that science was essential for handling practical problems, and agronomists and veterinarians participated in scientific meetings while scientists usually went into the field.

In 1932 Neiva left the institute and was replaced by Henrique Rocha Lima, known for his German education.[15] From the beginning, a natural division of labor was established between Neiva and Rocha Lima—the first more concerned with the external side of the institute (fund-raising, political contacts, sanitary campaigns) and the second as close as possible to the daily technical and scientific activities. When Neiva left, the institute had already established some of its main features as a serious scientific institution: full-time work for its researchers, interdisciplinary work, an excellent library, good technical support (photography, drawing, glass-making, publications), and two periodicals, the *Arquivos do Instituto Biológico* and *O Biológico*.

The tone of the institute appeared in its weekly scientific meetings. On Tuesdays internal meetings consisted of presentations and discussion of scientific articles in the recent literature led each week by a different researcher. On Fridays, lectures, often given by invited guests, covered broader scientific, literary, or artistic subjects and were open to the public. They soon became part of the city's intellectual calendar. It was not uncommon for Rio de Janeiro's scientists and intellectuals to travel to

13. Reis 1976a, 1976b, 1976d.
14. Bier interview.
15. Rocha Lima was the son of a prestigious medical doctor in Rio de Janeiro and had been in contact with the Manguinhos group since his student years. In 1901, after graduating from Rio's medical school, he went to Berlin, where he specialized in pathological anatomy, a new field for Brazil. He became a member of Manguinhos' staff in 1903, and in 1907 he left Brazil again to work at the University of Jena as an assistant professor of pathological anatomy at the invitation of a former professor, Hermann Duerk. His scientific career in Europe was considered brilliant and included a period in Hamburg's Institute of Tropical Medicine and in the local university. He reputedly made important contributions in research related to yellow fever and typhus, among other areas. He kept up an intense correspondence with Oswaldo Cruz, but returned to Brazil only when invited by Neiva to join the Instituto Biológico in 1927.

São Paulo for these meetings. Otto Bier, in his interview, stressed that these activities were very important in building the institute's prestige and recognition and helped to maintain what Neiva and Lima considered its "university spirit," which they deemed necessary to protect it from narrow specialization.[16]

Such a "spirit" was absent from Rio de Janeiro's Faculdade de Medicina, where Carlos Chagas Filho launched his Instituto de Biofísica (Biophysics Institute) in 1937. The Faculdade was unanimously described by its contemporaries as having inadequate equipment and poorly organized courses in which the best students were those who could attach themselves to a prestigious professor and practice in his infirmary or laboratory. The basic disciplines, which were supposed to provide future physicians with scientific knowledge, were the most neglected. There was only one exception, physiology—thanks to the teachings of Alvaro Osório de Almeida who provided students with an exciting image of what scientific research could be and invited some of them to his laboratory or to Manguinhos.

Chagas Filho entered the Faculdade de Medicina through a public competition for the chair of biological physics. His thesis was intensely discussed with Carneiro Felipe, chemist; Costa Ribeiro, physicist; and Antônio Oliveira Castro, from the Institute of Electric Technology of the Faculdade de Engenharia. It was the first time the Faculdade de Medicina had used physical and chemical methods in the study of biological phenomena. Once nominated to the chair, Chagas Filho left for study in France and England; on his return he began to organize his research institute at the Faculdade.

Personal and family contacts provided financial support to free Chagas and his collaborators from bureaucratic constraints. A position of "specialized technician" with a salary above that of an assistant professor was created in the Faculdade de Medicina through direct interference of

---

16. José Reis described the "university spirit" prevailing in the institute as "something one learns from great scientists and thinkers who are used to thinking in universal terms, interested in the exchange of ideas, and convinced there are no barriers among different fields of knowledge. It is a spirit of modesty, based on the acceptance of criticism and the never-ending need to learn. It is the spirit of open dialogue, not limited by differences of age or hierarchy but based on the respect of each other's personality and thought. It is the spirit of adventure in the search and transmission of knowledge, in which the intellectual, spiritual, and moral concerns always prevail over material concerns. It is the spirit of always starting anew" (Reis 1976a:593). The interviews with Penha, Bier, Rocha e Silva, and Reis, and the significant scientific production that came out of the institute throughout the years, confirm that this way of understanding their work was present in their minds and had a positive effect on their productivity.

Luís Simões Lopes, director of the powerful Departamento de Administração do Serviço Público. To his ability to get support, Chagas added a modern and updated view on how scientific work should be carried out: emphasis on interpersonal cooperation and exchange of information among all scientists; openness to curiosity, questioning, and exchange of ideas; curtailment of excessive bureaucratization and hierarchy. Chagas would say that a scientific institution is made first with people, then with problems and equipment, and finally with the workplace. This view was in contrast to the Brazilian tradition of starting with buildings, filling them with equipment, hiring personnel—and only then identifying research questions that could justify their existence.[17]

Chagas' laboratory quickly gained national and international reputation. He invited Tito Enéas Leme Lopes and Lafaiete Rodrigues Pereira, both trained in Manguinhos, and Oromar Moreira, José Moura Gonçalves, and José Batista Veiga Sales, all biochemists from Belo Horizonte endorsed by Baeta Viana. Herta Meyer, who used to work in the laboratory of pathology in Manguinhos, which was supported by the Rockefeller Foundation, started the laboratory of histology, together with João Machado. They conducted studies on the culture of protozoa, such as the *Trypanosoma cruzi* and the *Plasmodium aviarium*, with direct relevance to public health and were supported by the Serviço Especial de Grandes Endemias (Special Service of Large Epidemics), a nonofficial agency headed by Chagas' brother Evandro and supported with resources from the Guinle family; and the bioelectrogenesis of the hearth tissues, stemming from strictly academic motivations. The next laboratory to be organized was the one in biophysics headed by Chagas and with the cooperation of Bernhard Gross, from the Instituto Nacional de Tecnologia. The main subject was the bioelectrogenesis of the *Eletrophorus eletricus*, an electric fish found in the Amazon region that provided a unique vehicle for studying the interaction between biological and physical processes.

The comparison between these two institutions—the Instituto Biológico and the Instituto de Biofísica—shows some common elements and important differences. Both were led by strong personalities trained in Europe and in Manguinhos—Neiva, Rocha Lima, and Chagas Filho. Both benefited from the intense and close interpersonal links their leaders maintained with strong political figures, which were essential to protect them from the sameness forced on all institutions by the public bureaucracy. Finally, they shared a new and daring view of the role of modern science as strongly based on fundamental research and enlarged intellectual horizons.

17. Mariani 1982b.

That the Instituto Biológico was an institution for applied research would finally make a difference. For some years Rocha Lima's prestige and strong personality, combined with his family ties to the state-appointed governor, Fernando Costa, was enough to protect the institute from outside interference. After 1937, however, with Ademar de Barros in the state government, not only the Instituto Biológico but also all scientific and educational institutions in São Paulo began to suffer.[18] In 1949 Rocha Lima resigned, and the Biológico entered a period of slow decay.

In contrast, the best years of the Instituto de Biofísica were still to come. Chagas' leadership, which would continue until the 1980s; the protection against short-term demands and external interference; the relative independence from the professional courses; and a strong commitment to high standards of excellence all combined to make the Instituto de Biofísica a genuine heir of Manguinhos' best traditions.

Chemistry: Limits and Possibilities of the German Model.

Brazilian chemistry has always counted on significant German participation, as illustrated by a list of some of the field's most important names.[19] Theodore Teckolt, born in German Silesia and a University of Rostock pharmacy graduate, reorganized the Museu Nacional's chemistry laboratory in 1874 under Ladislau Neto; Wilhelm Michler, born in Württemberg, studied at Stuttgart Polytechnic Institute, received his doctorate in Zurich with Victor Meyer, and was appointed professor of industrial chemistry at the Escola Politécnica in Rio in 1884. He used private funds to organize a chemistry laboratory where he could carry out his work and train disciples: F. G. Dafert, who organized Campinas' Estação Agronômica in 1887; Alfred Schaeffer, who earned his bachelor's degree in pharmacy and his doctorate in chemistry from the University of Munich and organized not only Minas Gerais' Laboratório de Análise do Estado

---

18. Maurício Rocha e Silva recalled those years as a period of "complete disaster": "At one point they cut the scientists' salaries, ended full-time work, and created so many difficulties that many preferred to leave the research institutions and get a job in the private sector. The Instituto Butantã probably suffered the most, and for a while some of its scientists—for example, Anatol Rosenfeld and Leal Prado—took shelter at the Biológico" (Silva interview).

19. For the history of chemistry in Brazil, see Rheinboldt 1955 and Mathias 1975.

in Belo Horizonte in 1911 but also that of the Escola de Engenharia some years later; Otto Rothe, who had a doctorate in chemistry from the University of Jena and was hired in 1920 to set up the chemistry course at Porto Alegre's Escola de Engenharia and in 1926 succeeded Schaeffer in Belo Horizonte; and several others.

Participation by Germans became even more pronounced with the arrival of Heinrich Rheinboldt at the Universidade de São Paulo and of Fritz Feigl at Rio de Janeiro's Laboratório de Produção Mineral, which belonged to the Ministry of Agriculture. Rheinboldt was followed by his assistant, Heinrich Hauptmann, and later by Herbert Stettiner, Hans Stammreich, and Pawel Kromholz. Hans Zocher, formerly a professor at the universities of Berlin and Prague was to work with Fritz Feigl.

There were three reasons behind this strong German presence: the economic and migratory bonds linking Brazil and Germany until the 1930s; the problem that young German professors faced in the job market due to the traditional rigidity of the German universities; and the crises and persecutions caused by Germany's growing Nazi movement, which led to the withdrawal not only of Jewish scientists but also of those who were simply liberal. Brazil, on the other hand, was very interested in German chemistry, perhaps because of the German tradition of integration of chemical research and industrial activities. This is therefore an excellent testing ground for the possibilities of transplanting such a tradition to a different social and economic context. As we shall see, the failure to reproduce this integration contributed to the difficulties encountered in attempting to establish chemistry in Brazil.

The German presence was not exclusive. A detailed survey of professors of chemistry and authors of chemistry books, carried out by Heinrich Rheinboldt, shows a long list of non-German names in the country's schools of medicine and engineering, some of them rated as very original and competent.[20]

20. One of them was Alvaro Joaquim de Oliveira, a military engineer and author of *Apontamentos de Química*, a book Rheinboldt rates as "the best and most original Brazilian work" in the field. Rheinboldt points out that Alvaro de Oliveira was, with Benjamin Constant Botelho de Magalhães, one of the founders of the Sociedade Positivista, and it may have been for this reason that "he was led to defend the theory of the constancy of valence so unilaterally, which lent to his work a peculiar trait. The works of Alvaro de Oliveira deserve the attention of a qualified philosopher!" Rheinboldt's own position in relation to positivism in Brazil was one of cautious perplexity: "It is very peculiar that this doctrine—which clearly indicates what paths should be followed in future chemistry research and which led J. H. Van't Hoff, for example, to make marvelous discoveries—did not spark what was so necessary in Brazil: the abolition of the old cramming system and the birth of pure research. But not even Alvaro de Oliveira himself undertook a single original experiment" (Rheinboldt 1955:69).

Rio de Janeiro's Instituto de Química (Chemistry Institute) was the first institution specifically devoted to this subject. It was organized as a center of research and training that was to provide "strictly scientific courses for the training of professional chemists" and short courses for nonspecialists who would learn "certain aspects of applied chemistry for use in industry and commerce."[21] These short-lived courses eventually inspired the creation of various other industrial chemistry courses throughout the country. In 1920, funding of 100 "contos de réis" (approximately £6,000) was made available for courses to be established in Belém, Recife, Salvador, Belo Horizonte, Ouro Preto, Rio de Janeiro, São Paulo, and Porto Alegre. For a while this bill gave a boost to the chemistry studies in engineering courses at São Paulo's Escola Politécnica; at the engineering schools of Belo Horizonte (which hired Schaeffer and von Burgher), Porto Alegre (which hired Otto Rothe and E. Schirm), and Recife; and at Niterói's Escola Superior de Agricultura e Veterinária. Earlier, in 1926, the Escola Politécnica de São Paulo had merged its chemistry and industrial engineering courses to form its course in chemical engineering. When federal funding was cut in 1930, this experience practically ended. Only the institutions that had begun their chemistry programs independently continued to exist. In 1934 a new institution, the Escola Nacional de Química (National School of Chemistry), was created in Rio, but it never developed into a significant research institution.[22]

In hindsight it is clear that initial attempts to implant chemistry in Brazil failed because the country did not reproduce the special combination of a strong academic environment and an active chemical industry that was so striking in the German case. Experience showed that if

21. Excerpts from the decree that created this institute are cited in Mathias 1975:17.
22. First headed by Freitas Machado and later by Carneiro Felipe, it was linked to the Departamento Nacional da Produção Mineral, which maintained its industrial chemistry course until 1951, when it was transformed into a chemical engineering course. The Escola seemed to have been particularly closed to outside influences. From 1939 on, the Laboratório da Produção Mineral hired Fritz Feigl, of international reputation, who was joined in 1946 by Hans Zocher. Jacques Danon, who studied at the Escola Nacional de Química, recalls that they had no influence in the school because they were forbidden to teach. "The Brazilian scientific community—the community of professors, to be more precise—was extremely jealous of the privileges derived from their chairs, and it was afraid of more creative people. I don't blame them; I understand their social conditions. The presence of such important names threatened those who posed as scholars but lacked creativity" (Danon interview). The Escola Nacional de Química went through several transformations in the years to follow, but the teaching of chemistry as an independent discipline was institutionalized in Rio de Janeiro only when the Instituto de Química of the Universidade Federal do Rio de Janeiro was established in the 1970s.

neither of the two conditions existed chemical research would benefit more from emphasis on academic work than from emphasis on applied results.[23]

Chemistry research at the Universidade de São Paulo was established in the German tradition by Heinrich Rheinboldt, who arrived in Brazil in 1934 as an established scientist.[24] He was accompanied by Heinrich Hauptmann, who obtained his doctoral degree in chemistry under Fritz Strauss and had worked in Göttingen with Adolf Windaus, who received the Nobel Prize for chemistry in 1928 for the discovery of cholesterol; and by Herbert Stettiner, who had obtained his doctorate from the University of Berlin in 1928.

The chemistry department of the Faculdade de Filosofia came to be known for the practical, systematic, and empirical orientation Rheinboldt and his team gave it. "The first groups of students were very small. I belonged to the third or fourth generation, and we were twenty-four students. We lived in the laboratories, from eight in the morning to six in the afternoon. We left the laboratories only to attend classes."[25] Rheinboldt took care of theoretical and experimental classes and the teaching of general and inorganic chemistry and analytical chemistry, with the assistance of Stettiner. Hauptmann was in charge of practical work and the teaching of physical, organic, and biological chemistry.[26]

Rheinboldt always referred to his department as "the institute," in the German tradition, and behaved accordingly. In addition to the license in chemistry, granted for secondary school teachers, he also offered a doctoral degree for those who could write a dissertation based on an original research project under the direction of a professor. Two of the first four students, Simão Mathias and Pascoal Senise, got their doctoral degrees and were hired by the department.[27]

At first, the chemistry and physics departments were housed at the

---

23. For Simão Mathias, a student of Rheinboldt's at the Universidade de São Paulo, the industrial chemistry courses of the 1920s failed because they were "merely for professional training, not aimed at profound neutral studies or at original research" (Mathias 1975:21). The latter was to be the goal of the Universidade de São Paulo's chemistry department.

24. Rheinboldt was born in Baden, graduated with degrees in chemistry and geology from the Technical Institute of Karlsruhe, and had a doctorate from Strasbourg under the direction of W. Wedekind. In 1927 he was already head of the department of analytical and inorganic chemistry at the University of Strasbourg, and in 1928 he taught as *extraordinarius* at the Institute of Chemistry in Bonn.

25. Mors interview.
26. Mathias 1975:11.
27. It is believed that Mathias received the first doctoral degree awarded by the Universidade de São Paulo.

Escola Politécnica, but when they were not well received they moved to the pharmacy section of the Faculdade de Medicina. That was worse,[28] and sometime later the department was housed in a provisional building at the Alameda Glete.

Hostility from the professional schools did not stop the chemistry department from training good chemists. Those who graduated from the Faculdade de Filosofia were considered the best generation of professional chemists ever educated in Brazil and had no difficulty finding jobs in the growing number of national and multinational industries being established in São Paulo and elsewhere. This does not mean that there was any kind of interaction between the department and some agency of economic or industrial planning, or with the private sector. Except for the war years, when the department helped with the development of quartz crystals for the sonar project, there was almost no direct contact between their professors and the industry or governmental agencies.

This was contrary to the German experience and therefore considered a bad thing. Simão Mathias described the situation:

> If we look at the important chemistry departments in Germany and elsewhere after World War II (or even after World War I in the United States), we see that they had intense contacts with industries. There were always contracts and other forms of collaboration between the industrialists and the scientists. This is an old German tradition that was taken up by the Americans. Here [in Brazil] unfortunately, it was never understood. Our laws of full-time dedication forbid such arrangements. Nobody ever favored the contacts between industry and science in our country. When I was director of the chemistry department [at the Universidade de São Paulo] I made several approaches to the Federação das Indústrias looking for some kind of integration. Nothing came out of it.

The problem was not limited to lack of understanding or rigid regulations for full-time work.

---

28. "When the building of a new chemistry laboratory was under way," Mathias recalls, "the medical students staged a protest: 'We don't want philosophers in the Faculdade de Medicina!' For them we were the philosophers from the Faculdade de Filosofia. One night fire was set at the scaffolds. It was the end of the chemistry department at the Faculdade de Medicina" (Mathias interview).

> Most of our chemical industries are controlled by multinational corporations or associated with them. These industries have their laboratories in their countries of origin and are not interested in doing scientific research in Brazil. . . . The country needs to create its own technology, relevant to our reality, instead of transferring technology from more advanced countries and forcing on us a system developed by them.

Not only did the industrial companies turn their back on the chemistry departments, but so did the government: "We never had well-defined applied research projects. No problem relevant for the country was ever brought to us to research. One example is alcohol. Now the government has woken up and discovered that alcohol could replace oil. We chemists have known it for centuries." This isolation, however, did not seem to have hurt the quality of research carried out at the Universidade de São Paulo. "Chemical phenomena have no boundaries. They happen here as they do on other planets, as we are learning now with astrophysics. All our scientific work is being published in international journals and adds to the existing knowledge in this scientific field."[29]

In other words, the chemical researchers at the Faculdade de Filosofia longed for a much more applied work, but their relative success[30] is explained by the fact that they, willingly or not, geared their work toward basic research. The misconceptions built into Mathias' frustrations with his department were well captured by Joseph Ben-David in a report he wrote about the Brazilian scientific community and its frustration over applied work after a brief visit in 1976.

> Limiting research and training to the requirements of such technologically defined problems would be in the long run highly

---

29. Mathias interview.
30. The professional competence and high standards developed by Rheinboldt and his group are undisputed. However, they may not have been as up-to-date in their discipline as their physicist colleagues were. Mathias recalls that Hauptmann's course in physical chemistry at the Universidade de São Paulo was "a disaster." Paulus A. Pompéia, a physicist, says that Rheinboldt and Hauptmann "were great chemists, but of the nineteenth century": "The Germans had gone a long way in classical chemistry, but they did not know physics, did not know quantum mechanics, did not know the physical part of chemistry. I believe this was a problem peculiar to Germany, because in other places the chemist worked very closely with the physicists" (Mathias interview). This is probably the reason several talented young men who came to study chemistry at the Universidade de São Paulo, such as José Israel Vargas, did not find the intellectual answers they were looking for and soon moved to the physics department.

inefficient. The people trained for such purposes would have great difficulty learning new technologies, and research of such limited kind would become obsolete in a short time. New technological needs would require new plans for training and research, and the maturation of the plans would usually lag far behind the needs. . . . Contrary to the myth that developing countries cannot afford pure science and must adjust their investments in research and training to precise economic goals, this would be for them the most unreasonable thing to do. Since they have great uncertainties about the future course of their economic-technological developments, by subjecting their efforts to narrow considerations they are likely to misdirect and waste them.[31]

Rheinboldt and Hauptmann continued in Brazil with the research interests they had defined in Germany. Rheinboldt's work was related to the "study of organic and molecular compounds of sulfur, and later the organic compounds of selenium and tellurium." Hauptmann worked on the composition of natural products found in Brazil, including the chemistry of coffee. The research line on natural products was carried on by Walter B. Mors, who studied under Hauptmann and in 1943 joined a newly created Instituto Agronômico do Norte in the city of Belém, one of several agricultural research stations then established by the Brazilian Ministry of Agriculture. The institute worked, among other things, on natural rubber, a product that had military significance in those years; that research received strong support from the U.S. government. Mors worked on the properties of a plant known locally as "timbó," which produced a substance used in the production of insecticides and of strategic relevance as well. Later he helped organize the Instituto de Química Agrícola (Institute of Agricultural Chemistry) in Rio de Janeiro, also a branch of the Ministry of Agriculture but dismantled by administrative decree in 1962.[32]

31. Ben-David 1976:17–18.
32. A strong research group on the chemistry of natural products was developed at the Instituto de Química Agrícola with the cooperation of Carl Djerassi, from Wayne State University and later Stanford, who previously had been a leader of the research arm of Syntex Corporation. (Syntex held the patents for the production of hormones for birth control pills derived from Mexican cacti.) After 1962 Otto Gottlieb, a member of the group, went to organize the chemistry department of the new Universidade de Brasília, while Mors began a research center of natural products at the Faculdade de Farmácia in Rio de Janeiro, later incorporated in the chemistry department of the Universidade do Rio de Janeiro.

Gleb Wataghin and Cosmic Ray Physics

Modern physics started in Brazil at the Faculdade de Filosofia, Ciências e Letras of the Universidade de São Paulo through particle physics and, unlike chemistry or the biological sciences, without any practical applications or results. In the following decades there would be several chances to prove its practical worth, and physics as a discipline would become the strongest research field in Brazil.[33]

Physics research began with Gleb Wataghin, who had been invited by Teodoro Ramos through Enrico Fermi. Along with Francesco Cerelli from the Italian Academy of Sciences, Fermi helped locate suitable candidates. Wataghin was one of them, the other being mathematician Luigi Fantappié, then twenty-nine years old.[34] Wataghin did not belong to Europe's first rank of physical scientists, but he was close enough to know the leading names, understand their work, and identify suitable research questions for himself and his students.[35] In 1927 he attended an international physics conference in Como, where he became acquainted with the best-known physicists of his time. In 1930 he published an article in the *Zeitschrift für Physik* having to do with nuclear particles and forces, which was discussed at the Solvay conference of that year and led to an exchange of letters with Enrico Fermi. In 1931 Wataghin began his studies on cosmic rays (a research line began in 1921 by Robert A. Millikan in the United States and continued by Arthur Compton) and published his first article on the so-called relativistic cut-off. In 1933 Wataghin traveled through Europe, spending a few months in contact with Lord Rutherford in Cambridge and a few weeks in Co-

---

33. For a detailed account of the development of modern physics in Brazil, see R. G. F. Pinto 1978.

34. Cerelli had previously been in Brazil and had discussed with Armando de Sales Oliveira the projects for the new university. Wataghin heard that Fermi had suggested his name and reacted negatively at first. Then Teodoro Ramos came and invited him "to Rome, where we went to a famous restaurant—Via de la Scrofa—where the spaghetti was eaten with spoons and forks of pure gold" (Wataghin interview). He finally conceded.

35. Wataghin was born in Odessa and did all his secondary studies in Russia. His father was an officer and engineer in the Russian army, and the whole family migrated to Italy in 1919, after the revolution. In Turin, Wataghin did translations from Russian to Esperanto, taught Latin and mathematics, and worked in the film industry. In 1922 he obtained a doctorate in physics from the University of Turin, and in 1924 he was hired as an assistant by the polytechnic school of that university. Five years later he received from the Italian Ministry of Education the *libera docenza* in theoretical physics to teach rational mechanics and advanced physics.

penhagen with Bohr, experiences that would mark his entire life as a scientist.[36]

Wataghin and Fantappié at first shared a small office on the third floor of the Escola Politécnica. "We had to give a complete course. Fantappié taught all mathematical subjects. I taught experimental and theoretical physics and theoretical mechanics. We gave many classes. Besides, I was told that I should start an experimental laboratory. I have always preferred theory; I could, however, start with cosmic rays, high energies. I could use a laboratory for that."[37] His first students in the Politécnica included Mário Schenberg, Júlio Rabim, Cândido da Silva Dias, and Cavalcante Albuquerque. Between 1934 and 1942 he developed two research lines: one in theoretical physics, with Schenberg, Abraão de Morais, and Walter Schutzer; the other on cosmic rays, with Marcelo Damy de Souza Santos, Paulus A. Pompéia, and Yolande Monteux.

Wataghin was mainly a theoretical physicist.[38] Between 1934 and 1936 he published several theoretical papers dealing with statistical mechanics of light particles in high temperatures and with relativistic quantum electrodynamics. His most distinguished student in theoretical physics was probably Mário Schenberg. Schenberg had arrived in São Paulo from Recife, where he had studied with Luís Freire. It soon became clear

---

36. "I recall especially two types of events from my time in Cambridge. On two or three Sundays I was invited for tea at Rutherford's house. Everybody came. I met Geiger and became a friend of Dirac's. These gatherings gave me a glimpse of English society, which in those years was usually so exclusive. There were not only scientists but also ladies. For me, these meetings were extremely interesting and useful." The other events were the weekly meetings of the so-called Kapitza Club. "Kapitza was a soviet citizen and worked closely with Rutherford. He is four or five years older than me, which means he was about thirty-six or thirty-seven at that time.... We became friends—we are both Russian—and we used to play chess. I think he won most of the time, but it did not matter. The important thing was the friendship, the conversations...." The next stop was Copenhagen. "For the first time I met personally with Bohr. There was also Heitler, Heisenberg, Pauli.... Bohr invited me to present my ideas. Pauli chaired the meeting. Everybody was against me because I believed cosmic rays came from multiple sources." From Copenhagen he went to Leipzig, where Heisenberg worked in a period of great excitement. "There I met Jordan, Debye, Max Born—who was just arriving—and Ettore Majorana, a very young man who impressed me as a true genius, which he really was" (Wataghin interview). Wataghin was unknown except for his paper on the Solvay conference, and he was always impressed by the informality and cordiality with which he was received in this small elite.

37. Wataghin interview.

38. Marcelo Damy observes, however, that "Wataghin was a theoretical physicist with a strong interest in the experimental side. He knew very well that theory had to be based on facts because physics is a natural science.... But he was not an experimentalist; he was not a man to design equipment, to make it, to adjust it for observation. His contribution was in planning experiments and analyzing the results" (Damy interview).

that he had an extraordinary talent for mathematics and physics, and in 1936 he went to Rome to work with Fermi for two years.[39] In 1939 Schenberg was invited by George Gamov, who had been to Brazil, to work with him at George Washington University. Together they developed a theory in astrophysics that became known as the "Urca process," in a reference to the Urca Casino in Rio de Janeiro. Later he spent some time at Princeton's Center for Advanced Study and in the Yerkes Astronomic Observatory with Subrahmanyan Chandrasekhar, returning to Brazil in 1942. In 1944 he received the chair of rational mechanics of the Faculdade de Filosofia of the Universidade de São Paulo.

Marcelo Damy, who had left engineering for physics, was Wataghin's main assistant on the experimental side since his graduation in 1937.[40] Research work intensified in 1938 with the arrival of Giuseppe Occhialini, who had studied with Patrick M. Blackett in Cambridge. According to Damy, he brought to Brazil the tradition of experimental physics then being developed with J. J. Thompson and Lord Rutherford, and a new series of experiments with cosmic rays was begun. At the end of 1938 Damy received a fellowship from the British government to go to Cambridge. "In Cambridge I worked with Sir Henry Bragg and his son William Lawrence Bragg, who like his father was a Nobel prize winner. William Bragg was a specialist in X-rays, and I had another research supervisor, Professor H. Carmichael. Carmichael was Cambridge's specialist in cosmic rays and worked with Walter Heitler and H. Bhaba, both Nobel prize winners and very famous."[41] With the war, the Cambridge scientists became involved in the research to develop radar, and Damy was invited to stay and join the group since the new technology required very precise measurements. There was official contact between the British Foreign Office and the Brazilian Foreign Ministry about this possibility, but the Brazilian government did not go along. In 1940 Damy returned to Brazil.

---

39. Wataghin recalls: "Schenberg returned a different person. He had learned much more than I could have taught him. From then on we collaborated. He did beautiful work on cosmic rays and then started to work in electrodynamics, under the direction of Dirac. He had learned a lot in Rome, and I decided that he did not have much more to learn from me and should travel again soon" (Wataghin interview).

40. "I started to work on problems related to cosmic rays, which required very special technologies. For instance, all observations were done with equipment based on electronic circuits. Radiation was detected with the famous Geiger-Müller counters, then not very well known. But there were no electronic circuits or radiation detectors on the market. The physicist had to design and make his own circuits, build the detectors with his hands, and then use them in his research" (Damy interview).

41. Damy interview.

In early 1939 Damy was replaced as Wataghin's assistant by Paulus A. Pompéia, who had graduated from São Paulo's Politécnica in 1935.[42] Wataghin, Occhialini, and Pompéia launched a series of studies on cosmic rays, using airplanes from the Brazilian air force flying up to seven kilometers high. At the invitation of Arthur Compton, Pompéia went to the United States in 1940 and worked for two years under the supervision of Norman Wilberg (later director of Chicago's Argon Laboratory) developing new measurement techniques and electronic circuitry.

In 1941 Compton organized an expedition to South America to measure the impact of cosmic rays in the Bolivian Andes and in the São Paulo region. Pompéia returned to Brazil to join Wataghin and Damy in preparing for the event. They worked with stratospheric balloons, and Wataghin demonstrated the existence of "penetrating showers" of cosmic rays, which were evidence of multiple production of mesons. In 1942 Compton became the director of the Metallurgical Laboratory, which was working on the atomic bomb, and Pompéia returned to Brazil. With all major scientific groups in England and the United States involved in the war effort, Wataghin and his group were for a while the only ones working on cosmic rays. Very shortly, however, the Brazilians also would get involved in military technology.

The War Effort

"One or two months after my return from England," recalls Damy, "Paulus Pompéia and I were approached by the Brazilian navy about the possibility of developing equipment for the detection of submarines.... We had many ships torpedoed by German and Italian submarines and did not have any equipment to detect them. Although Brazil had joined the Allies in the war, it not only did not receive the new sonars and radars, but Brazilians were not even allowed to come close to these top secret military installations. Before that, we had also been approached by the army. They were making their own cannon balls, with Brazilian-made powder, and needed to develop ways to measure the speed of bullets. This was our first military assignment."[43]

In both cases the physicists were approached after Brazilian authorities realized that no one else in Brazil could carry on with such projects.

---

42. Pompéia had worked from 1935 to 1938 as an assistant to Fonseca Teles in the Instituto de Eletrotécnica, where he organized a laboratory for physical measurements.
43. Damy interview.

When we got this assignment from the navy we made clear to the person in charge, Admiral Guilherme Bastos Pereira das Neves, that we lacked experience with naval problems and had no experience with submarine detection, that we were just "philosophers" working with cosmic rays. But to study problems of basic science we were led to use nonconventional methodologies to demonstrate the existence of certain phenomena. We were therefore used to facing and dealing with the unknown. We believed that, at least from a psychological point of view, we had the proper attitude to approach the problem. Besides, we did not think it would be too difficult. There were available a reasonable number of publications on techniques used for submarine detection in World War I. Our problem was not to discover new laws of nature but to rediscover, so to speak, the conditions under which a beam of ultrasound could be emitted and detected and how to measure the time interval to identify the submarine's position.[44]

For the army, Pompéia developed an instrument that could measure the initial speed of bullets with an accuracy of 0.4 percent.[45] He and Damy also developed portable radios for the army's jeeps and trucks. The most interesting projects, however, were those for the navy. The first product was an instrument that could hear the sound of the submarine's propellers. Later they developed equipment to send an ultrasound beam but were unable to capture its echo. Still later they developed a complete sonar. In its final version, the ultrasound transmitter was made out of 400 cylinders of nickel welded into a base of steel that had to turn continuously. The echo was captured by a crystal detector. A special problem was the production of quartz crystals of adequate size, a problem handed over to and solved by the chemistry department of the Faculdade de Filosofia. A special thermostat based on the dilation of gasoline was developed by Damy and Pompéia to control the crystal's cooling temperature.

The development of such equipment, which was completely unknown in Brazil at the time, required that a series of new technical problems be solved by incorporating new specialists and institutions—such as the

44. Damy interview.
45. "Measurements of small time intervals were completely unknown to the Brazilian engineers—that I knew because I had worked in the United States with measurements of half-lives of mesons, which had the magnitude of microseconds. It was a very specialized technology and very new. . . . We built equipment that measured the time the bullet took to cross two light beams" (Pompéia interview).

Liceu de Artes e Ofícios, the Instituto de Pesquisas Tecnológicas, and the Instituto de Electrotécnica, all in São Paulo. In all, eighty sonars were built for the navy; twenty-two industries supplied parts for the equipment, without knowing their final destination. The sonars themselves were assembled in the building of the Faculdade de Filosofia at Avenue Brigadeiro Luís Antônio by Damy, Pompéia, and eighteen specialized technicians.

After the war, the navy ended its cooperation with the physicists of the Universidade de São Paulo, who returned to their academic and scientific chores. The technological know-how the group developed had spilled over to other institutions and private companies that began work on the production of electric equipment and other sophisticated products for the postwar consumer market. As the country's economy opened up, these industries were with a few exceptions either displaced by foreign imports or acquired by foreign corporations getting established in Brazil.[46]

Postwar Developments

The war brought Fantappié back to Italy, but not Wataghin, who became an expatriate from his country of adoption. For the Brazilians, Wataghin was Italian enough to be kept out of the war projects being developed by Damy and Pompéia. He had to step down as head of the physics department at the Faculdade de Filosofia, but he continued with his research projects with the assistance of Oscar Sala and Elza Gomide. Sala entered the university in 1942 and was immediately called to help with the Compton expedition.

Sala recalls that when Wataghin invited him to participate in his cosmic rays studies the physics department at the Universidade de São Paulo had already stopped all its academic research projects. His first problem was to rebuild all the equipment, "more sophisticated than that used before by Damy and Pompéia."[47] First Wataghin and Sala put their equipment in the ceiling of the Faculdade de Medicina; later they moved to a garage in a small hotel in the mountain resort of Campos de Jordão. They worked with few resources and little support.[48]

    46. Leff 1968.
    47. Sala interview.
    48. "There was no money to pay for my expenses in Campos de Jordão or for transportation. Wataghin would pay the expenses out of his pocket, or we would look for well-

After the war, with Damy as its director of the physics department, the Rockefeller Foundation donated $75,000 for acquisition of a particle accelerator in the United States for the physics department. Damy and Wataghin went to the United States to choose the equipment and decided on a 23-megawatt betatron. For a year Damy remained in Illinois working with Donald W. Kerst on the equipment.[49] Upon his graduation in 1945, Sala was invited to work as an assistant to Damy, and he went to Illinois in 1946 with Paulo Bittencourt, also with the support of the Rockefeller Foundation. There he worked with Maurice Goldhaber in neutron physics. In 1948 he went to Wisconsin to prepare for the acquisition of a new piece of equipment for the Universidade de São Paulo: a Van der Graaf electrostatic accelerator.

The research line developed by Wataghin on cosmic rays was continued by Cesare Lattes, who studied at the Faculdade de Filosofia between 1941 and 1943. In 1943 he worked with Occhialini, who had also remained in Brazil during the war. In 1944 the university hired him as the third assistant for the chair of theoretical and mathematical physics, and he later became involved in experimental research. In 1945 Occhialini, who had moved to the University of Bristol the year before, invited Lattes to join him. In Bristol, Lattes worked under the general direction of Cecil Powell with Blackett, Conversi, Pancini, and others. He was reputedly responsible for planning the experiments that led to the discovery of the meson-pi, predicted by Hideki Yukawa some years earlier and for which Cecil Powell received a Nobel prize. After Bristol, Lattes was invited to Copenhagen to present his results; then he went to work at Berkeley's cyclotron with George Gardner, "with the clear intention of trying to obtain the artificial production of heavy mesons, since the light mesons should be produced by their disintegration."[50] There, he was able to produce the meson-pi and show how it disintegrated into the

---

known wealthy people and ask for their support. Once Wataghin went to talk with São Paulo's governor, Ademar de Barros. Wataghin was an enthusiastic person, and in that conversation—I was not there, I heard the story later—the governor got very impressed, opened a drawer, took a pack of bills and asked, 'Professor, how much do you need?' It is a funny story, which reveals how Wataghin viewed the governor" (Sala interview). It also shows Wataghin's isolation in those years.

49. The installation of the betatron presented an opportunity to train a new group of scientists, including José Goldemberg, R. Pieroni, and others. "It was the first equipment and allowed for the beginning of nuclear physics in Brazil" (Damy interview).

50. Lattes interview.

meson-mi and a new particle, the neutrino.[51] Lattes returned to Brazil in 1949 to organize the Centro Brasileiro de Pesquisas Físicas in Rio de Janeiro, of which he was the first scientific director.

That same year, Wataghin returned to Italy to become director of the physics institute at the University of Turin. From then on he would return to Brazil only on special occasions: in 1952 for a brief course; in 1955 to receive the degree of doctor *honoris causa* from the Universidade de São Paulo; and in 1971, when the physics institute of the new Universidade de Campinas was given his name.

While these developments were taking place in São Paulo, a different research tradition was being established in Rio de Janeiro, even though on a smaller scale, through the works of Bernhard Gross and Joaquim Costa Ribeiro.

Gross was born in Germany, studied engineering and got his doctorate at the Technical Institute in Stuttgart, where he did research on cosmic rays. He came to Brazil in 1933 and met Dulcídio Pereira, who held the chair of physics at the Escola Politécnica and was assisted by Joaquim Costa Ribeiro, Francisco Mendes de Oliveira Castro, and Eugênio Hime. He was invited to present his work at the Politécnica and in the recently created Instituto Nacional de Tecnologia, where he began to work.[52] His assignments were technical and applied, but he managed to start his own research projects.

> In 1934 the local electric company wanted to measure the electric resistance of telephone cables, and of their insulation. We began with the measurements. The cables presented a phenomenon that had fascinated me in Germany, together with the cosmic rays. It was what was called di-electric absorption. We began to measure them with very rough instruments. As the work progressed, we also began theoretical studies, and the papers we produced then

---

51. Lattes' work in Berkeley is described by José Leite Lopes: "An important achievement in physics, the discovery of pions and of the disintegration pion-muon, and the work of Marcello Conversi and associates in Italy, on the capture of mesons in cosmic radiation, marked the birth of particle physics as an independent field from nuclear physics, after the years of limited scientific achievements during World War II" (Lopes 1988:2).

52. "I was completely alone in a room that had . . . , well, in an empty room. They still did not have a physics division. Actually, they had it on paper, and the director was Aníbal de Souza, who later moved to the department of industrial property. He did not do any work of physics at the institute; he was more interested in patents and such things. In the beginning I borrowed some electric equipment from the Observatório Nacional that had been purchased by Henrique Morize. I needed a source of high tension, and a battery of 500 volts was obtained. There was also a galvanometer. I have no idea on how we got it" (Gross interview). With these instruments, Gross began work.

are in a way still valid because this subject is as relevant today as it was then.[53]

In 1937 Gross became director of the newly created division of standards of the institute, which was to define legally valid standards for weights and measures. As a German, he was replaced as director in 1942 by Oliveira Castro. During the war Gross had a minor role in the war effort, which included development of a clockwork mechanism for the detonation of grenades. Caught by circumstances, he continued with his research interests, and in 1942 he identified a phenomenon he called the "freezing" of electric currents in electrets. His work from 1942 to 1945 was published in three articles in the *Journal of Applied Physics* (1947, 1948, 1949). He continued to work at the Instituto Nacional de Tecnologia in different capacities after the war.

Gross' main associate, Joaquim Costa Ribeiro, graduated in engineering from the Politécnica in Rio de Janeiro and became *livre docente* in 1933. He held the chair of experimental physics at the Universidade do Distrito Federal, while Gross held the chair of general physics and had Plínio Sussekind da Rocha as his assistant. When the Universidade do Distrito Federal was closed in 1939, the whole group moved to the Faculdade Nacional de Filosofia. Costa Ribeiro cooperated with Gross in different projects related to electrets and in 1942 demonstrated the existence of a "thermal-dielectric effect" that became known as "Costa Ribeiro effect." In 1946 Costa Ribeiro took the chair of physics at the Faculdade Nacional de Filosofia, where he continued with his research and formed a significant group of students, including Paulo Saraiva de Toledo, Armando Dias Tavares, E. Rodrigues, and Sérgio Mascarenhas. Mascarenhas, in turn, organized the group of solid-state physics at the Universidade de São Carlos in the state of São Paulo, where Gross worked in his later years.

Some broad generalizations can be made about these experiences. First, the most successful developments, and those more amenable to practical utilization in the long run, were those with a stronger academic orientation. Second, they all benefited from the presence of foreign visitors or immigrants—Wataghin, Rheinboldt, Brieger, and Dobzhansky—who knew how to form disciples and create traditions of research work. Third, they sent their best students to international research centers at an early stage.

These experiences saw achievements but also problems and failures,

53. Gross interview.

some of which have already been discussed and others of which are yet to be seen. In any event, they provided Brazil, and more specifically the Universidade de São Paulo, with a scientific density that no other educational institution in the country had ever had. The scientists coming from this university, along with those from the Manguinhos tradition, would form the basis for the developments that would take place after World War II. It is impossible to follow these developments in all their details, but a broad view of their direction and a discussion on present and future dilemmas is the subject of the second part of this book.

PART TWO

GROWTH

7

TOWARD A SCIENTIFIC ROLE

Scientific institutions and communities with stability and space for long-term projects and growth can prosper only when society comes to recognize and accept science as a profession. Some of this recognition existed in the old museums and other scientific institutions of imperial times, but none survived the pragmatic mood of the Republic. When the protective emperor left, those who wanted to do research had to teach in professional schools, work on sanitary campaigns, produce medications, take care of patients, work on engineering projects, and look for mineral riches. Long-term scientific work could be done only during leisure time or hidden under applied activities. It was not only that society did not recognize science but also that the scientists themselves, with a few exceptions, lacked a clear view of their role and place in society. In the following decades this role began to take shape in the scientists' minds, and they worked to obtain recognition from society, a struggle that is far from complete.

We have seen different aspects of this process in the preceding chapters. Here, we look at it again through a broad comparison between different generations of scientists that shaped the Brazilian scientific community, which should bring the previous chapters into clear focus. As a reference, we shall take a closer look at the professional careers of the sample of fifty-six leading scientists interviewed for our study.[1] The dividing line between generations is always arbitrary, but the sample falls naturally into three groups. The first contains those who were born at the turn of the century, whom I call the pioneers; the second group is made up of those who were born ten to twenty years later, who were introduced to the sciences by the first and who created the first modern scientific institutions in Brazil; in the third group are scientists who studied in these new institutions and are the bridge between the older generation and today.

The Pioneers

With few exceptions, the Brazilians in the oldest generation studied either engineering or medicine in Rio de Janeiro. They were all children of

---

1. This was hardly an unbiased sample. There was a preference for older people, who could give a personal account of events further back into the century. The age bracket includes people born between 1892 and 1931. (We excluded from this analysis the younger ones who were also interviewed.) There is also a biological bias, since we could interview only those who were alive in 1977. Finally, we were concerned mostly with the biological and hard sciences, so mathematics, engineering, medicine, the social sciences, and the humanities were excluded. The tables in this chapter were first published in Schwartzman 1984a, and their limitations were revealed by the comments they provoked (Cruz 1985; Ladosky 1985; Mors 1985; M. da S. Pinto 1985). Some people who should have been included in the study were left out for good or poor reasons, and the opposite may be also true. Any complete listing in biological sciences should include entomologist Angelo da Costa Lima; zoologists Ernst and Evenine Marcus; botanists Frederico Carlos Höhne, Felix Rawitscher, and Mário Guimarães Ferri in São Paulo; João Geraldo Khulman, Kurt Brade, and Carlos Toledo Rizini in Rio de Janeiro; and Adolfo Ducke and João Murça Pires in Belém. A significant group of followers of Carlos Chagas Filho should also be included, starting with Aristides Pacheco Leão. Frota Moreira, the younger (Table 4), was part of this group and would probably best be placed in the next generation. Gottlieb and Mors (Table 5), although slightly older, see themselves in the same generation as Wladislaw and Giesbrecht. Mors began his career as a researcher only in the late 1940s in the Instituto de Química Agrícola in Rio de Janeiro, and Gottlieb also had a late start. The younger generation would be best represented by the inclusion of Haiti Moussatché, Walter Oswaldo Cruz, Johana Döbereiner, Moura Gonçalves, and Wilson Beraldo. However, even such an improved listing would still be incomplete.

middle-class, educated families. Their parents were small-businessmen, physicians, and teachers, so contact with some kind of intellectual activity was not alien to them.

Of the four names in the physical sciences (see Table 1), only the first, Lélio Gama, had, properly speaking, a scientific career. He graduated with a degree in engineering from the Politécnica in Rio de Janeiro and later participated in the group of mathematicians headed by Otto de Alencar and Amoroso Costa. Lélio Gama worked as an astronomer in Rio de Janeiro's Observatório Nacional with Henrique Morize.

Another member of this group, Francisco Magalhães Gomes, taught physics at the Escola de Minas and later at the Escola de Engenharia de Minas Gerais, but was never a researcher in the proper sense of the word. He was influential in orienting a small group of outstanding scientists who got their training in São Paulo and abroad. The other two, Othon Leonardos and Mário da Silva Pinto, were mostly men of action and were involved in creating institutions set up by the Brazilian government to exploit Brazil's natural resources. Leonardos was also the author of a significant work on the history of Brazilian earth sciences.

Silva Pinto does not see himself as among Brazil's true pioneers in geology, "such as Glycon de Paiva or Otávio Barbosa." He describes himself mostly as a "technologist, manager, and specialist in raw materials." He considers his work on applied geology, economic geography, hydrography, and navigation secondary and incidental. After World War II, however, he helped organize the teaching of geology and establish geology as a profession. He also helped negotiate cooperation agreements between Brazil and the U.S. Bureau of Mines and the U.S. Geological Survey.[2]

By comparison, the biologists of this generation tended to be much more defined in their scientific roles (see Table 2). They all studied medicine, and almost all went through the Instituto Manguinhos in Rio de Janeiro and further advanced training abroad. They had an older generation to provide a pattern, including Oswaldo Cruz, Adolfo Lutz, and Ezequiel Dias. São Paulo had also its own research group in bacteriological and tropical diseases, but in contrast to Manguinhos it did not develop a research tradition of its own and did not survive for long. It gave rise, however, to the Instituto Butantã, where Afrânio do Amaral made his scientific and institutional career.

Both the physical and biological disciplines were French-oriented, but the field of tropical medicine and public health came under North American influence quite early, carried on mostly through the Rockefel-

2. M. da S. Pinto 1985.

TABLE 1. Physicists and Geologists, First Generation (1892–1907), First Degrees in Brazil

Year of Birth and Name	Specialization and Education	Place of Birth and Family Background
1892 Lélio Gama	Astronomer and mathematician, Politécnica do Rio de Janeiro	Rio de Janeiro, father a military engineer
1899 Othon Leonardos	Geologist, Politécnica do Rio de Janeiro	Minas Gerais, father a businessman
1906 Francisco Magalhães Gomes	Physicist, Escola de Minas de Ouro Preto and Universidade de Minas Gerais	Minas Gerais, father a professor at the Faculdade de Medicina
1907 Mário da Silva Pinto	Geologist and metallurgist, Politécnica do Rio de Janeiro, Departamento Nacional de Produção Mineral	Rio de Janeiro, father a professor at medical school, mother a schoolteacher

TABLE 2. Biologists, First Generation (1892–1907), First Degrees in Brazil

Year of Birth and Name	Specialization and Education	Place of Birth and Family Background
1894 Afrânio do Amaral	Tropical medicine, Faculdade de Medicina do Rio de Janeiro, Harvard University	Pará, Brazil, father an entrepreneur (rubber plantation owner)
1895 Olímpio da Fonseca	Parasitologist, Faculdade de Medicina do Rio de Janeiro, Manguinhos, U.S., and France	Rio de Janeiro, father a medical doctor
1904 Adolfo Martins Penha	Faculdade de Medicina de Minas Gerais, Manguinhos	Interior of Minas Gerais, parents died early
1905 Otto Bier	Bacteriologist and immunologist, Faculdade de Medicina do Rio de Janeiro and Manguinhos	Rio de Janeiro, son of European immigrants
1907 José Reis	Bacteriologist, Faculdade de Medicina do Rio de Janeiro, Manguinhos, Instituto Biológico de São Paulo, and Rockefeller Institute	Rio de Janeiro, father a small businessman
1907 Amílcar Viana Martins	Zoologist, Faculdade de Medicina de Minas Gerais and Rocky Mountain, U.S.	Minas Gerais, father a public employee

ler Foundation, which was present in Brazil in 1916. It worked both directly, in campaigns against yellow fever and ancylostomiasis, and through institutional support to the Faculdade de Medicina de São Paulo, which from the beginning adopted several features of the American system of medical education, including full-time work for professors and *numerus clausus* for students. There was close cooperation between the Rockefeller health specialists and Manguinhos, and several Brazilians of the next generation would continue their training in the United States through this channel.

In short, the first generation of Brazilian scientists was trained in conventional courses of engineering and medicine. For some peculiar reason its members were related to one of the few places in the country where some kind of scientific concern existed—the Observatório Nacional, the Instituto Manguinhos, or its counterpart in São Paulo. A scientific role, however limited, already existed for the biologists at the time, but not for the others. We see little or no continuation coming out of the old imperial institutions, such as the Museu Nacional or the Jardim Botânico.

It is interesting to contrast these pioneers to the group of foreign-born Brazilians in the same age bracket who arrived in the 1930s and had a great influence on the country's scientific environment (see Table 3). They were all born at the turn of the century and had their training not in the liberal professions but in science as such. They came with their doctorates already completed, and some had already begun an academic and scientific career. The reasons they came vary: some were displaced by the prewar tensions in Europe; others were not happy with their career prospects; others were still very young and adventurous and accepted a long stay in a far-away country. Several of the Italians came with support of the Mussolini government in what was considered an important cultural mission from Italy to Brazil. The same thing happened with some of the Frenchmen, whose government had an active policy of cultural dissemination.[3]

Only a few of those who came remained in Brazil and continued their academic life. The achievements of those who did stay can be explained not only by the fact that they had better training than the Brazilians and knew firsthand what science was about but also because, coming later, they could benefit from a much better though still quite limited institutional environment. Since their professional identities were already established, they could use them to shape the institutions they helped organize. Wataghin and Brieger were part of the first group of professors of the

---

3. Pyenson 1982 and 1984.

TABLE 3. Scientists Educated Abroad, First Generation (1892–1907)

Year of Birth and Name	Education and Specialization	Place of Birth, Family Background, Year of Arrival
1889 Gleb Wataghin	Physicist, Turin, Italy	Russia, father an engineer, arrived in São Paulo 1934
1900 F. Brieger	Geneticist, University of Breslau, Germany	Germany, father a physician and professor, arrived in São Paulo 1934
1902 Quintino Mingóia	Chemist, University of Pavia, Italy	Italy, arrived in São Paulo 1935
1903 Guido Beck	Physicist and mathematician, Vienna, Cavendish Laboratory, Leipzig and other places	Arrived in Rio de Janeiro 1951
1904 Viktor Leinz	Geologist, University of Heidelberg	Arrived in Rio de Janeiro 1933
1905 Bernhard Gross	Physicist, Stuttgart and Electric Research Association, London	Arrived in Rio de Janeiro 1933

Faculdade de Filosofia in São Paulo. Mingóia came in 1935 with a contract to work in a private institution, the Laboratório Paulista de Biologia, and in 1945 was hired as a professor at the Faculdade de Farmácia of the Universidade de São Paulo.[4] Gross helped establish the Instituto Nacional de Tecnologia in Rio de Janeiro, participated in the creation of the Universidade do Distrito Federal, and with Joaquim da Costa Ribeiro started the teaching of physics in the Universidade do Brasil. Guido Beck came later, and his influence was also significant.

Second Generation: Beginnings of Professionalization

Scientists of the second generation all had similar careers, following very closely the road opened by the first. Those in the biological sciences, almost without exception, graduated from the schools of medicine in Rio de Janeiro and São Paulo and later had access to the Instituto Manguinhos in Rio or the Instituto Biológico in São Paulo (see Table 4). The pattern is similar: while in medical school the young student caught the attention of a professor who worked also in Manguinhos, and started his apprenticeship. The Instituto Biológico was established in São Paulo in 1927 and was directed by Arthur Neiva, of Manguinhos' first generation, and all its initial staff was trained in Rio de Janeiro.

There was little in terms of scientific knowledge or scientific models that a student could get at the faculties in those years. Carlos Chagas Filho reported that the Faculdade de Medicina in Rio de Janeiro in the 1920s had "no practical courses, no seminars, no contacts between professors and students; only professorial lectures, given with punctuality and great eloquence."[5] Because of these conditions, scientists usually held the professional schools in contempt and limited their contacts to lectures and the eventual recruiting of some gifted students. To meet a prestigious scientist and work under his guidance was the only way for a young student to start a scientific career. Family ties helped: Walter and Oswaldo Cruz Filho were sons of Oswaldo Cruz; Evandro and Carlos Chagas Filho were sons of Carlos Chagas; Emanuel Dias was a son of Ezequiel Dias. For others, introduction to a scientist came through a

---

4. Mors 1985.
5. Chagas Filho interview.

family friend. That was how Olímpio da Fonseca Filho, Otto Bier, and José Reis, among others, started their careers.[6]

Once in touch with a patron, the next step was to begin working in his laboratory outside the university. Besides the few public institutions, there were also private initiatives, the most famous being the laboratory kept by the brothers Alvaro and Miguel Osório de Almeida in the basement of their home in Rio de Janeiro, which Chagas Filho describes as the place where research in physiology started in Brazil.[7]

Manguinhos' training course (curso de aplicação), launched in 1909, was the first organized path to a scientific career in the country. Admittance was possible only by invitation, and instruction was provided informally in a system of internship without formal courses or lectures. The trainees had to learn procedures for sterilization and handling glass, tasks usually given to laboratory assistants. Later the course became more formal, and lectures on microbiology were given for eighteen months. New modifications were introduced in 1913–14 with the inauguration of the institute's permanent building: the course became more rigid and formalized and lasted for fourteen months, there was a strict system of exams and evaluation, and those who missed ten classes were dropped from the course. About twenty trainees were selected each year, but only one-third to one-half usually made it to the end.[8]

A few other, less organized, alternatives existed. The Museu Nacional admitted "voluntary assistants" by invitation only. After a year the voluntary assistant could be promoted to nonpaying trainee. Then, in the unusual event of an opening, the trainee could be hired as a naturalist—

6. Bier and Reis interviews.
7. "Thanks to the help of Cândido Gaffrée, an associate of Eduardo Guinle in the organization of the Companhia Docas de Santos [a private corporation that controlled the docks of the Santos harbor in São Paulo], Alvaro Osório put together his small laboratory, . . . which became a cultural center that attracted intellectuals of all origins, including Amoroso Costa, the founder of modern mathematics among us. Miguel Osório was trained there. Silva Melo and Tales Martins, among others, participated in informal gatherings." Such an environment could stimulate intellectual curiosity but could hardly provide for a continuing career, as revealed by the fate of Miguel Osório de Almeida. "Miguel Osório, who had an exceptional intelligence, was a victim of the limitations of the Brazilian scientific environment. He was defeated in a public competition for the chair of biological physics, when he gave an extraordinary demonstration of culture and arrogance. . . . He did not know with whom to speak. Very close to the French school, he was lost in an endless stream of correspondence, letters, and long voyages, but always restricted to the Sorbonne, although I am sure that with his ability to work, his intelligence, and his culture he could have an extraordinary impact in another environment" (Chagas Filho interview).
8. Fonseca 1974:13–14.

TABLE 4. Biologists, Second Generation (1908–1920), First Degrees in Brazil

Year of Birth and Name	Specialization and Education	Place of Birth and Family Background
1908 José Ribeiro do Vale	Biochemist, Faculdade de Medicina de São Paulo, and U.S.	Minas Gerais, father a farmer
1909 Hugo de Souza Lopes	Entomologist, Escola de Agricultura e Veterinária, Rio de Janeiro	Rio de Janeiro
1910 Zeferino Vaz	Geneticist Faculdade de Medicina de São Paulo and Instituto Biológico de São Paulo	São Paulo, father a businessman
1910 Maurício Rocha e Silva	Biochemist, Faculdade de Medicina do Rio de Janeiro, Instituto Biológico de São Paulo, U.S., and England	Rio de Janeiro, father a liberal professional
1911 Carlos Chagas Filho	Biophysicist, Faculdade de Medicina do Rio de Janeiro, Instituto Manguinhos and University of Paris	Rio de Janeiro, son of biologist Carlos Chagas
1911 Herman Lent	Entomologist, Faculdade de Medicina do Rio de Janeiro and Instituto Manguinhos	Rio de Janeiro, father a small businessman
1914 Wladimir Lobato Paraense	Parasitologist, Faculdade de Medicina do Pará and Pernambuco, and Faculdade de Medicina do Rio de Janeiro	Rio de Janeiro

TABLE 4 (*continued*)

1914 Mário Viana Dias	Neurophysiologist, Faculdade de Medicina do Rio de Janeiro and National Institute of Medical Research, U.S.	Rio de Janeiro, several physicians in the family
1919 Crodowaldo Pavan	Geneticist, Faculdade de Filosofia, USP and Columbia University, U.S.	São Paulo, father an entrepreneur
1920 Manuel da Frota Moreira	Physicologist, Faculdade de Medicina do Rio de Janeiro and studies in the U.S. and England	Rio de Janeiro, father a physician

a general term encompassing a large variety of subjects from ethnology and ethnography to mineralogy or petrography and including botany, zoology, and linguistics.

This was also the period in which a few people, even those not professional scientists themselves, played a crucial role in spreading scientific values, finding talents, and stimulating their scientific careers. Baeta Viana, in Belo Horizonte, is often presented as an example. He was less a scholar or specialist than a propagandizer for a new approach to medical science. He graduated from the medical school in Belo Horizonte and was one of the first Brazilians to go to the United States with a fellowship from the Rockefeller Foundation. He worked in the States for two years, in a period of rapid expansion of biochemistry, and on his return to Brazil he came into direct conflict with the French tradition that dominated the school. Still, he managed to carve a place for himself and organized one of the best medical libraries in the country. His students, including Moura Gonçalves and Wilson Beraldo, would be counted among the best biochemistry specialists in Brazil.[9] André Dreyfus, one of the founders of São Paulo's Faculdade de Filosofia, played a similar role in a very different context.[10] Men like Baeta Viana and Dreyfus therefore made the transition between old professor—highly rhetorical, bookish, self-sufficient, sometimes knowledgeable but insensitive and prejudiced against empirical work—and the modern scientist, trained to identify a problem, define it, and solve it.

Second Generation: The Hard Sciences

As the biologists started with medicine and then went to Manguinhos, most of those in the hard sciences started in the school of engineering

---

9. "It is difficult to find a good Brazilian biochemist who is not related, directly or indirectly, to the Baeta Viana school. And this is still more remarkable because by himself he was not a great researcher. No single important research work can be attributed to him" (Chagas Filho interview; Ladosky 1985).

10. Crodowaldo Pavan recalls how his career as a geneticist was set after he attended a conference by Dreyfus. "He was an incredibly stimulating teacher. Complicated things became very simple after his explanations. He could get to the core of a problem, explain it, and make everybody feel it made sense, even if one did not understand the question completely. His lectures were important events for young intellectuals. He would link genetics with histology and give classes and courses on psychoanalysis. When he became a full-time teacher at the Faculdade de Filosofia, he realized that all his time was being spent on lectures, something he really enjoyed. But his scientific basis, his experimental basis, was quite restricted for the kind of program he wanted to conduct (Pavan interview).

and moved to the Faculdade de Filosofia of the Universidade de São Paulo (see Table 5). Data on family background is scant, but they suggest that hard scientists came from less privileged families than the biologists. Although medicine and engineering had similar social standing at the time, Manguinhos was a prestigious institution, and its scientists intermingled easily with the country's elites. As for the Faculdade de Filosofia in São Paulo, it was elegant and becoming to a Paulista intellectual in the 1930s to attend lectures by foreign professors in the new institution, but making that their career was another matter altogether.

The Escola Politécnica was similar to the Faculdade de Medicina in its shortcomings regarding scientific research. Gross recalls that the flame of physics was kept alive only by the lectures of one teacher, Dulcídio Pereira.[11] In the 1920s, under Eusébio de Oliveira, the Serviço Geológico initiated its own system of internship.[12] But the most important road to science was through personal influence. What Dreyfus did for the biological sciences, and more, Luís Freire in Pernambuco did for physics and mathematics. Freire's list of students is impressive: physicists Mário Schenberg, José Leite Lopes, Fernando de Sousa Barros, and Ricardo Ferreira; mathematician Leopoldo Nachbin. One of them recalls:

> Freire was very stimulating but never became a scientist himself. He was a very competent and brilliant teacher, very stimulating, but he was not a person who could guide and form his students.... He was a scholar of the type we can find in all Latin countries. They are extremely knowledgeable professors who get

11. Maurício Rocha e Silva relates his experience: "I wanted to become a physicist before studying medicine, and I used to come to the laboratory of Dulcídio Pereira in the Politécnica.... I got a terrible impression from the laboratory. It was worse than in medicine; they did absolutely nothing.... There was a spectrometer, new for those years, which only Dulcídio Pereira could touch. Everything else was for teaching physics at the gymnasium. Anyone who wanted could use it, but nobody did.... There was nothing in theoretical physics. The only famous mathematician, was Amoroso Costa, with whom I liked to get in touch.... All the others were technicians pretending to be mathematicians" (Silva interview).

12. Mário da Silva Pinto recalls how the system worked in the Laboratório da Produção Mineral: "We wanted to be as close as possible to the institutions of higher education—in chemistry, close to the Escola Nacional de Química; in metallurgy, close to the school of engineering and with Ouro Preto. The students were admitted as trainees without pay in the first year. We had organized a true learning program, and they had to go through all sections of the laboratory, from the preparation of samples to the processing of minerals, and through the sections of physical chemistry and chemistry. We selected those with more aptitude for production and invited them to take an examination for selecting those who would remain for another year as paid interns. Later they could apply through a public examination and begin their professional career" (Pinto interview).

TABLE 5. Physical and Chemical Scientists, Second Generation (1908–1920), First Degrees in Brazil

Year of Birth and Name	Specialization and Education	Place of Birth and Family Background
1908 Simão Mathias	Chemist, Faculdade de Filosofia, USP and University of Wisconsin, U.S.	Father a small businessman
1909 Paulus A. Pompéia	Engineer, Politécnica de São Paulo and University of Chicago	São Paulo, father an engineer
1914 Mário Schenberg	Engineer and physicist, Faculdade de Filosofia, USP; Italy and U.S.	Pernambuco, father a European immigrant
1914 Marcelo Damy de S. Santos	Physicist, Universidade de São Paulo and Cambridge, England	São Paulo
1917 Pascoal A. Senise	Chemist, Universidade de São Paulo, and Louisiana State University, U.S.	São Paulo, son of Italian immigrants
1918 José Leite Lopes	Chemist, Universidade de Pernambuco; physicist, Universidade de São Paulo; Princeton University	Pernambuco, father a small businessman
1920 Walter B. Mors	Chemist, Universidade de São Paulo, and University of Michigan	São Paulo, family of immigrants
1920 Otto Gottlieb	Chemist, Escola Nacional de Química, Rio de Janeiro, England, and Israel	Czechoslovakia, secondary education in Europe; arrived in Brazil with his family before World War II
1920 Jaime Tiomno	Physicist, Universidade do Distrito Federal and Faculdade de Filosofia da Universidade do Brasil; Princeton	Rio de Janeiro, son of immigrant, small businessman

the latest publications and have an incredible personal library in their homes. They know everything, give beautiful lectures, and could teach in any university in the world. But they are not scientists; they do not come down to carry on a limited research work. Freire was a good example of that. He was born in Recife in 1900, studied engineering, became a professor of physics, and wrote a few articles that were published in the *Annales de la Physique* in France, but I believe he never became a scientist because of the circumstances in which he lived.[13]

José Leite Lopes was less enthusiastic, but he confirms Freire's influence.[14]

Modern Scientists: The Third Generation

The novelty with members of the third generation was that for the first time in Brazil they had the chance to go directly to a science course without going through a professional school first. Those who were not in São Paulo attended some of the short-lived chemistry courses in their regions and then transferred to the Universidade de São Paulo or went abroad. During World War II and afterward, the Rockefeller Foundation began to provide fellowships for Brazilian scientists outside the health sciences, which benefited many scholars of this generation.

The different sizes of Table 6 and Table 7 reflect a sampling bias and do not necessarily mean that the group of hard scientists was larger than the group of biologists. But they also reflect the fact that in the 1930s and 1940s physics was the most prestigious scientific discipline, and that in Brazil as in other places, physics attracted an unusual group of gifted minds. A look at the family backgrounds in Table 6 and Table 7 confirm that the hard scientists came from families that were far less established than the biologists. While most of the biologists kept to their professional work, many physicists joined the country's intelligentsia and became well-known public figures engaged in the general discussions on the role of science, technology, and education in Brazil's development. It is as if

13. Ricardo Ferreira interview. See also Mota and Hamburger 1988.
14. "Because of Freire, I began to study physics and mathematics more seriously. Of course he could not teach the same way as in Europe, or even like a specialist who is in contact with large scientific centers. Recife was a province in Brazil, but relatively speaking the professors there were capable of opening minds, attracting students, showing the way, and providing the main principles of the sciences" (Lopes interview).

TABLE 6. Physical and Chemical Scientists, Third Generation (1921–1931)

Year of Birth and Name	Specialization and Education	Place of Birth and Family Background
1921 Blanka Wladislaw	Chemist, Faculdade de Filosofia, USP	Poland, family arrived in Brazil in 1935
1921 Ernesto Giesbrecht	Chemist, Faculdade de Filosofia, USP	Father a civil engineer
1922 Oscar Sala	Physicist, Faculdade de Filosofia, USP, and Illinois and Wisconsin, U.S.	Italy, family of immigrants, did all his studies in Brazil
1923 Aluisio Pimenta	Pharmacist, Universidade de Minas Gerais	Minas Gerais, father had a pharmacy
1924 Jacques Danon	Chemist, Escola Nacional de Química, Rio de Janeiro, and Paris	Father a small businessman
1924 Cesare Lattes	Physicist, Faculdade de Filosofia, USP, and Princeton, U.S.	Paraná, family of Italian immigrants, father a bank clerk
1925 Paulo Leal Ferreira	Physicist, Faculdade de Filosofia, USP, and Rome	Rio de Janeiro, father an engineer
1925 Jean Meyer	Physicist, Faculdade de Filosofia, USP, and École Polytechnique, Paris	Danzig (Gdansk), secondary studies in Europe, family of immigrants
1926 Sérgio Porto	Chemist, Faculdade de Filosofia, USP, and physics, Johns Hopkins University and Bell Laboratories, U.S.	Niterói, Rio de Janeiro, father a small businessman
1928 Roberto Salmeron	Engineer, Faculdade de Engenharia, USP; Faculdade de Filosofia, Rio de Janeiro; and Manchester, England	Rio de Janeiro

TABLE 6 (continued)

Year	Name	Description	Origin
1928	José Israel Vargas	Chemist, Universidade de Minas Gerais; physicist, Universidade de São Paulo, and Cambridge University	Minas Gerais, father a small industrialist
1928	José Goldemberg	Physicist, Universidade de São Paulo, and studies in Canada	Rio de Janeiro, father an engineer
1928	Ricardo Ferreira	Chemist, Pernambuco, physicist, Universidade de São Paulo, and California Institute of Technology, U.S.	Pernambuco, father a small businessman
1930	Gerhard Jacob	Mathematician and physicist, Faculdade de Filosofia, Rio Grande do Sul	Germany, family of immigrants
1931	Rogério Cerqueira Leite	Engineer, Instituto Technológico da Aeronáutica, and physicist, Bell Laboratories, U.S.	São Paulo

TABLE 7. Biologists, Third Generation (1921–1931)

Year of Birth and Name	Specialization and Education	Place of Birth and Family Background
1922 Warwick Kerr	Geneticist, Escola Superior de Agricultura Luis de Queiroz	São Paulo, father a specialized worker
1923 Paulo Emílio Vanzolini	Zoologist, Universidade de São Paulo and Harvard University	São Paulo, father an engineer related to the Escola Politécnica in São Paulo
1925 Antônio Cordeiro	Geneticist, Faculdade de Filosofia, Universidade do Rio Grande do Sul, and Columbia University	Rio Grande do Sul, father a military officer
1928 Francisco M. Salzano	Geneticist, Faculdade de Filosofia, Rio Grande do Sul	Rio Grande do Sul, father a physician

the biologists tended to remain in their already established positions of social privilege while the physicists, in a clear drive for social mobility, took up a much more explicit intellectual role. In some ways they reproduced the European scientistic movements of the past, trying to grasp the most prestigious field of knowledge of their time and from that position acting to influence society as a whole.[15]

The Faculdade de Filosofia in São Paulo would introduce in Brazil new working standards that were almost unknown at the time. Marcelo Damy recalls the courses offered in 1934, when the introductory courses of the Escola Politécnica were combined with those of the new institution:

> I had the chance to follow the courses of mathematical analysis with Luigi Fantappié, geometry with Giacomo Albanese, and physics with Gleb Wataghin. We met a completely different world. In our education as future engineers, we still had the type of lectures so common in most of the Brazilian universities: the professor comes in, delivers his lecture, and walks away, without talking with the students and often teaching from an obsolete book. These professors were not researchers; they had other professions and taught only a few hours a week. For most, their own education was very deficient. There was strong inbreeding in the school, with one engineer training another for teaching basic disciplines. Because of that we believed disciplines like mathematics, chemistry, and physics were the study of things that were completely solved, crystallized, dead. For us, physics was something that resided in the books of physics; the same held for chemistry and mathematics. It was a surprise for us when we attended lectures that followed a completely different approach, that showed us these sciences were not only alive but going through such intense change that the amount of research published in the last few years had been greater than the amount since the beginnings of these sciences.... We also came in touch with something that was totally unknown in Brazil—the seminars. Each week the Italians and the Germans, who taught chemistry, would get together to present their research or the main lines of fundamental research being developed abroad. Then there was an open interchange of views. For us—young students used to listening without questioning—it was strange to hear a professor raise questions about and strongly criticize the work of a colleague. Very often the criticism was correct, but that did not mean the researchers would not remain

15. Schwartzman 1984a.

friends and that life would not continue as always. So we learned that science was alive. It could be developed, it was being developed in the rest of the world, and this possibility was opened also to Brazil.[16]

From the initial group of foreign professors a new model of scientist, which would have an extremely important role in the years to come, was built. The testimony of Gleb Wataghin allows us to see how this was done:

> I came from Italy with Fantappié. We received from the Faculdade de Filosofia one office, and we were told to teach. We asked for a library.... I was lucky. I found very able and interested young men without doing anything toward it. Who could assure a young man in 1934 that if he followed a course during three or four years he could become a professional physicist? Anyhow, they wanted to do science and I taught them what they wanted. Among them were Marcelo Damy de Sousa Santos, Mário Schenberg, and later Paulus A. Pompéia. In the Escola Politécnica, where I taught, I tried to tell the students that one could not do several different things at once. Then some of them decided to leave the engineering courses and dedicate themselves to physics. They knew about electricity, how to build radios, antennas.... Because of that, it was easy for them to work in experimental physics.... As much as possible I tried to send them to Europe after two or three years of study. I sent Mário Schenberg to my friend Dirac, who I believe is the most important physicist alive. I went to Europe with Schenberg; we went through Italy on the way to England. I met Fermi and asked him to talk with Schenberg. It was then that Fermi convinced Schenberg to work with him. I did the same thing with the experimental physicists. Some, like Lattes, went to Cambridge, England. They would write to me, showing solutions to technical problems—how to improve a circuit we had done here, for instance. I learned a lot from my students, and I trained them with the help of great physicists from all over Europe, Germany, England, and Italy.... Contact with Europe was essential. The only condition I had when I came here was that I wanted to spend two or three months each year in Europe. This was very good for me and for Brazil.[17]

16. Damy interview.
17. Wataghin interview.

The impact of foreign professors in the biological sciences was less pronounced, probably because there was already a much more developed tradition of research in these fields. Besides, the German professors of zoology and botany—Breslau, Marcus, and Rawitscher—belonged to well-established traditions of taxonomic research that were stronger than but not so different from what was already being done in Brazil. They did not have the appeal of novelty that came with physics. An important exception was Friedrich Brieger, who came to the Escola Superior de Agricultura Luiz de Queiroz and who, with Dreyfus, was responsible for the beginnings of genetics research in Brazil.

The creation of the Laboratório de Biofísica at the Faculdade de Medicina in Rio de Janeiro in 1937, under the leadership of Carlos Chagas, was a landmark in the introduction of organized research activities in Rio de Janeiro's academic institutions. For Chagas at the time, Manguinhos was a place plagued with low salaries and little administrative autonomy and lacking students to teach and stimulate the researchers to keep on working and studying. The university, with all its limitations, was perceived as holding greater promise. The laboratory, later called the Instituto de Biofísica, was one of the two main outgrowths of the Manguinhos tradition, along with the Instituto Biológico de São Paulo.

Sources of Financial Support

Hidden behind all sorts of applied activities, fighting for a space in a few higher education institutions, relying whenever possible on personal fortune or powerful friends and relatives, Brazilian scientists in the 1930s began to look for more legitimate and stable bases and sources of support. To ask how science is financed is in a way to ask how science is institutionalized and accepted as a legitimate activity in a society.

The old scientific institutions were supported by the federal and state governments, and witnesses tended to characterize researchers' salaries as "decent" and adequate for people willing to live a methodical and modest life. To this simple compensation was added the sense of privilege derived from work that would transform Brazil into a civilized country, free from ignorance and backwardness. Public health specialists firmly believed that most of Brazil's problems were caused by the population's poor health, and they saw their role as much broader than simply striving for medical progress.

This belief in a good cause helps explain why the directors of Manguinhos circumvented the rigidity of its budget to finance all kinds of activities not predicted in their statutes. For about thirty years the institute put profits from the sale of a veterinary vaccine into a fund that was used freely to establish outposts, finance scientific expeditions, hire specialists, or buy drugs and equipment that could not wait for the lengthy procedures of the federal bureaucracy. In 1938, however, the Institute became completely dependent on federal allocations because it was no longer allowed to produce the vaccine. The impact was not only material but also psychological, since it made it clear that Manguinhos no longer had its special place in Brazilian society. The withdrawal of this source of income helped accelerate the institute's decay. Only the laboratories that maintained an independent source of support could continue their work: the laboratories of helminthology (headed by Lauro Travassos) and hematology (headed by Walter Oswaldo Cruz), both linked to Evandro Chagas' Serviço Especial de Grandes Endemias.

Eduardo Guinle, who supported the Serviço Especial with private resources, is the most important name in Brazil's short history of scientific philanthropy. Late in the nineteenth century, two associates—Cândido Gaffrée and Eduardo Guinle—secured the contract for exploitation of the Santos harbor in the state of São Paulo for one hundred years. As the Brazilian economy moved to São Paulo, Santos became the country's busiest port, and Gaffrée and Guinle's company became its wealthiest enterprise. In 1906 Santos' Harbour Company asked Carlos Chagas, father of Evandro and Chagas Filho, to develop a program to end the malaria epidemics in Itatinga, a region in São Paulo where the company was building a hydroelectric dam. This was the beginning of a long history of association between the entrepreneurs and the scientists. In 1923 a Fundação Gaffrée-Guinle with philanthropic purposes was created, and Eduardo Guinle's son Guilherme carried this tradition a long way.

A volume published in 1958 by the Instituto de Biofísica in Rio de Janeiro in honor of Guilherme Guinle shows the breadth of his support. He helped maintain the laboratory of Alvaro and Miguel Osório de Almeida and with Carlos Chagas created an international center for leprology, created and supported the Serviço Especial de Grandes Endemias, and provided resources for Chagas Filho's Instituto de Biofísica and several laboratories in Manguinhos.[18]

---

18. Instituto de Biofísica 1951. In the book, Paulo de Goes describes Guinle as "some kind of private research council." Walter Oswaldo Cruz thanks Guinle for being "he who allowed us to free science from petty restrictions of an obsolete bureaucracy, who made it possible for us to buy without constraints the equipment needed for our work, who protected us from capricious administrators, and who gave us the chance to do science cheerfully."

Assis Chateaubriand, for several years the owner of Brazil's largest network of newspapers and radio stations, was also known to help young scientists. A colorful and unpredictable man, the support he provided tended to be full of surprises. Lobato Paraense had a typical story about fellowships he and three other colleagues received from Chateaubriand to come to São Paulo from Recife. They arrived in Rio by ship with money borrowed from a professor and went to see their sponsor, Chateaubriand, at his *O Jornal* newspaper office. At first Chateaubriand did not recall sponsoring them, but later he took the phone and cajoled four wealthy businessmen to support the students for the next year. In only a few minutes they had their fellowships and could start their careers.[19]

Manuel Frota Moreira, a biologist from the Chagas group who was the key person behind the Brazilian National Research Council in the 1950s and 1960s, explains the Brazilian scientists' failure to get stronger support in those years:

> Scientific activity was considered a cultural activity, and few people, in Brazil or elsewhere, believed that scientific research could be an instrument for power, wealth, and development. The contribution of scientific research and scientific knowledge to economic and military power is a novelty that was recognized only after the atomic bomb was produced with knowledge derived from basic and pure research. Although we had many examples of how scientific research, scientific knowledge, and technology could be useful for development of a country, it is striking that it was seldom considered as such.[20]

Thus, to get support, scientists had to prove their practical worth. In 1935 Arthur Neiva organized a short-lived Diretoria Geral de Pesquisas Científicas within the Ministry of Agriculture, Industry, and Commerce, which was supposed to bring together institutions like the Instituto Nacional de Tecnologia, the Laboratório Nacional da Produção Mineral, an institute of meteorology, and an institute of animal biology to be headed by Alvaro Osório de Almeida. Its practical orientation was obvious. This Diretoria would have been the first federal agency directly responsible for scientific activities in the country, but it never really got off the ground; after a conflict between Fonseca Costa, director of the Instituto de Tecnologia, and the minister of agriculture, the institute moved to another ministry and the whole project was abandoned. The

19. Paraense interview.
20. Frota Moreira interview.

idea of scientific planning was already capturing minds, and in 1938 Chagas went to Paris to learn about the Centre National de la Recherche Scientifique headed by Jean Perrin and established by the Curies in the years of the Popular Front. Chagas took all the documentation related to the CNRS to Minister of Education Gustavo Capanema, who was "extremely interested," as Chagas recalls it, but completely unable to raise the interest of President Vargas. Only much later, in 1951, would a national research center be created.

The Rockefeller Foundation in Brazil

The third source of support for science, other than the government and the private sector, was international foundations, of which the most important for many years was the Rockefeller Foundation. The Rockefeller Foundation was established in 1909 as a philanthropic institution "to foster civilization, to spread knowledge, and to reduce suffering"[21] with a $50 million endowment from the Standard Oil Company of New Jersey. It was preceded by three institutions: the Rockefeller Institute for Medical Research established in 1901; the General Education Board of 1903, aimed at developing the natural sciences, agriculture, humanities, and arts, mostly in the North American southern states; and the Rockefeller Sanitary Commission of 1909, geared toward the control of ancylostomiasis in the U.S. southern states.

The achievements of the Sanitary Commission influenced the decision to turn the activities of the new foundation in the direction of the field of medicine and public health and extend it to other countries. International activities were institutionalized in 1916 with the establishment of the International Health Board (previously the International Health Commission), which was responsible for taking to other countries the work of eradication of ancylostomiasis, for establishing public health agencies, and for spreading modern scientific medical practices. Other epidemics, such as malaria and yellow fever, were also targeted. Another program was directed toward improving medical education and public health in the United States and abroad with fellowships and institutional grants.

At the end of World War I the Rockefeller Foundation and the Education Board established a program to support medical schools in Latin America, the Middle East, and Southeast Asia. Two commissions were

21. Shaplen 1964:6.

sent to Latin America in 1916: one to study the spreading of yellow fever and its sources of contamination and to make suggestions for its eradication; the other to identify centers for medical education and public health to support.

The commissions went to Ecuador, Peru, Colombia, Venezuela, and Brazil, and in that same year an agreement was drawn up with the Faculdade de Medicina de São Paulo, then directed by Arnaldo Vieira de Carvalho: two new chairs were to be established at the Faculdade, to be supported jointly by the foundation and the São Paulo authorities for five years. Two American professors—Oscar Klotz and Robert Lambert—came to teach in the new chair of pathological anatomy, and two others, S. T. Darling and Wilson Smilie, came to teach hygiene. Two Brazilian physicians, Geraldo Horácio de Paula Souza and Borges Vieira, were sent to study at the Johns Hopkins School of Hygiene and Public Health. The foundation required a system of full-time teaching and *numerus clausus* for admittance of students, and the Faculdade had to change its regulations and submit them for approval to the state authorities to adjust to these requirements.

Besides the two chairs, the foundation supported the construction of laboratories of anatomy, physiology, chemistry, pathology, and hygiene, while the state government agreed to build a hospital. Later a grant was provided for the construction of a new building. An Instituto de Higiene was established in 1924, and in 1945 it was transformed into the autonomous Faculdade de Higiene e Saúde Pública.[22] Other institutions received smaller grants. The Faculdade de Medicina in Minas Gerais received support to create its chair of pathology, and Carlos Pinheiro Chagas was given a fellowship to study in the United States and take up the chair on his return. He was supposed to be the first Brazilian fellow of the Rockefeller Foundation.[23]

There was also an agreement between the Brazilian government and the Rockefeller Foundation for eradication of yellow fever and ancylostomiasis. In a period of five years, twenty-five stations were to be created in eleven states. Resources would come from the states, the cities, the local landowners, and the International Health Board. In 1917 a service for prevention of ancylostomiasis was established within the Departamento Nacional de Saúde Pública and began to provide technicians, physical facilities, and transportation, while the International Board supplied medication and microscopists. At the end of 1924 a network of 122 stations in twenty states was already in place. Research on ancylostomia-

22. A. de A. Prado 1958:790, 794–95.
23. Braga interview.

sis was carried on at the Faculdade de Medicina in São Paulo in cooperation with the International Health Board.

An outgrowth of this campaign was the creation of local health services in several states to improve the medical and sanitary conditions of rural populations—first in São Paulo and Minas Gerais and later in other regions. The teams, made up of a doctor, a nurse, a sanitary inspector, and an administrative assistant, had to inspect the local sanitary conditions, do laboratory tests, treat ancylostomiasis, and give inoculations.

All the agreements between the Rockefeller Foundation and the Brazilian government were made through Carlos Chagas' father, who was also director of public health and director of the Instituto Oswaldo Cruz, formerly Manguinhos. Bowman C. Crowell, a pathologist from Bellevue Hospital in New York City, came to guide the work of Manguinhos' pathologists, including Magarinos Torres, César Guerreiro, Osvino Pena, and Carlos Burle de Figueiredo. This cooperation became still more intense with the yellow fever epidemics of 1928.[24] A laboratory for yellow fever research was established in the Instituto Oswaldo Cruz in 1937 with the support of the Rockefeller Foundation, which could work only on problems directly related to the disease.[25]

Another epidemic was malaria, transmitted by the mosquito *Anopheles gambiae*. In 1937 it was first identified in the interior of Ceará, in the Brazilian Northeast; in 1938 Evandro Chagas and his team of the Serviço Especial de Grandes Epidemias found that the whole rural population in the Ceará valleys was contaminated. In a short period they counted 14,000 deaths and more than 100,000 cases in the Jaguaribe Valley. In October 1938 a team from the Serviço de Febre Amarela made up of technicians from the International Health Board and Brazilians arrived in Ceará to start the campaign. In January 1939, the Brazilian government created the Serviço de Malária do Nordeste, which signed immediately an agreement of cooperation with the Rockefeller Foundation.[26]

During the 1930s the Rockefeller Foundation broadened its range of concerns, providing resources for basic research, graduate education,

---

24. Fonseca Filho 1974:73. The campaign against yellow fever had begun in 1923, and for that purpose the country was divided into two regions: the North, to be handled directly by the Rockefeller Foundation from its seat in Salvador, and the South, under the responsibility of the Departamento Nacional de Saúde Pública and the Instituto Oswaldo Cruz. In the early 1930s it was possible to achieve reasonable levels of control, but in 1932 a wild form of yellow fever was identified—which led to the notion that the carrier, the mosquito *Aedes aegypti*, could not be totally eliminated, making inoculation the only alternative available for the exposed populations (Braga interview).

25. Dias interview.

26. Soper and Wilson 1943:84–86; Picaluga, Torres & Costa 1977:79.

institutional development, and higher education. For this it worked through five divisions: international health, which carried on the international sanitary campaigns; medical sciences, dealing mostly with physiology, industrial medicine, and psychiatry; natural sciences, for physics and biology; social sciences, concerned with the fields of international relations, economics, and public administration; and arts and humanities, with emphasis on archaeology and classic culture.[27]

The activities of the foundation in Latin America were at first restricted to the field of health, but during World War II one of its European specialists, Harry M. Miller, was sent to Latin America to identify promising scientists in all fields of knowledge to receive the foundation's support. Candidates had to be intellectually qualified and related to an institution that could support them upon their return. Resources were also provided to purchase equipment and for visiting professors from abroad. The Faculdade de Filosofia of the Universidade de São Paulo was a main recipient of this support, and mostly in the fields of genetics, physics, and chemistry.

Harry Miller, a biologist by education, proved to be good at finding young talent, and the support the Rockefeller Foundation provided was of great importance for the third generation of Brazilian scientists. However, support for basic science never became the main concern of the Rockefeller Foundation in Brazil (Table 8). All resources went to public health until World War II, when the Escola de Sociologia e Política in São Paulo and the natural sciences received limited support. An increase in support for agriculture and medicine followed. The foundation increased its support in the years of the Alliance for Progress (between 1955 and 1960) and then retreated. In 1970 it supported only the Universidade da Bahia, Salvador, in the field of applied social sciences.

In short, beyond its direct contribution to the control of tropical diseases, the Rockefeller Foundation had a major impact in the Brazilian scientific community by exporting American expertise and institutional models and by giving a significant group of Brazilians direct exposure to the American scientific and educational milieu. More specifically, it was instrumental in replacing France with the United States as the place where Brazilian scientists go to get their education, inspiration, and models.

One might wonder whether, in the long run, this influence had a positive or a negative impact on the Brazilian scientific community. The relevance and benefits of the sanitary campaigns, from a humanitarian point of view, seem beyond dispute. More debatable, eventually, was the adoption of the American model of medical education and professional

27. Nielsen 1972.

TABLE 8. Rockefeller Foundation Contributions to Science, Research, and Education in Brazil, 1932–1975 (Thousand Dollars)

Period	Public Health	Medicine	Natural Sciences	Social Sciences	Agriculture	Others	Total
1930–35	1,719	—	—	—	—	—	1,719
1936–40	1,117	—	—	—	—	—	1,117
1941–45	634	—	6	10	—	10	661
1946–50	392	—	80	5	18	40	537
1951–55	49	224	84	—	265	76	699
1956–60	190	1,466	592	286	955	144	3,634
1961–65	127	411	419	8	345	49	1,365
1966–70	319	37	235	2	168	—	611
1971–75	11	—	—	450	—	—	462

SOURCE: Calculated from the Rockefeller Foundation's annual reports (V. M. C. Pereira 1978).

organization, which was introduced when the Faculdade de Medicina in São Paulo was reorganized and which became the standard for the entire country.[28]

In principle, it would have been possible to adopt a different model of medical education and health care without sacrificing the knowledge that could be obtained through contacts with the more advanced scientific centers. In practice, however, such a path would have required an awareness of alternate models and a strong commitment to one of them by Brazilian authorities. In their absence, the American pattern was simply copied as closely as possible as the model to be followed. This happened not only in medicine but also in other scientific fields, and as the United States evolved into the world's major scientific center, the adoption of American models became a standard practice for most Brazilian scientific institutions.

Centralized Administration and Scientific Research

These movements toward institutionalization were to suffer the impact of a general tendency to political and administrative centralization that gathered speed in the 1930s and hit the new scientific and educational institutions particularly hard. There were many who looked favorably on

28. Pena 1977.

the process of centralizing and homogenizing Brazil's educational and administrative systems, considering such moves to be signs of modernization. Fernando de Azevedo, who was asked to write the introduction to Brazil's 1940 census, knew both sides of the fence and tried to strike a balance

> to agglomerate, bring together, and strengthen the similarities of the federated states in the spirit of national Brazilian communion—this was the main task the government instituted under the new political system, beginning with strengthening the authority of the central power, expanding borders, eliminating local differences, and merging of rural and urban states and communities into one nation. Unifying educational systems—not by adopting identical teaching structures but by adopting the same basic guidelines or in other words by organizing public education according to a general policy and joint plans—is one way (certainly the most powerful and efficacious way) the new regime intended to attain national assimilation and reconstruction.[29]

The fact is that the strengthening of the central government and the attempts to place the state bureaucracy under the aegis of rational management and "scientific administration" had the unintended consequence of putting much of the scientific research that still existed in the country's capital into disarray, without leaving much in its place. In 1937 the Departamento Nacional do Serviço Público (National Department of Civil Service) was established under Luís Simões Lopes, a close adviser to Getúlio Vargas, with the task of bringing all Brazil's public administration under control. For the first time, such ideas as the merit system, professionalization, careers, technical training of civil servants, and utilization of scientific methods in administration were brought to Brazil. The assumption was that state *dirigisme* would only increase in the years to come and that it required a strong, centralized, and scientifically minded public service. The larger ambitions never materialized, but the impact of the department in the daily life of Brazilian public institutions was long-lasting.[30]

One of the first acts of the new department was to decide that public servants would no longer be allowed to hold more than one civil job. This decree, known as the law of "desacumulação," had an immediate effect

---

29. F. de Azevedo 1963:689–90.
30. Schwartzman (ed.) 1983: chap. 1, pp. 15–70; Daland 1967.

on the teaching and research environments.[31] Most scientists decided to leave their academic appointments and remain in their institutes, where payment was higher and it was possible to carry on research or technical work. Full-time employment was almost unknown in Brazilian higher education institutions at the time, except at the Faculdade de Medicina in São Paulo, which received support from the Rockefeller Foundation and was supposed to follow the patterns then being introduced at American medical schools. Leinz recalls that his salary at the Departamento Nacional da Produção Mineral, about three "contos de réis," was ten times that of a tenured professor at the Escola de Engenharia. The indiscriminate application of the "desacumulação" law to teaching and research activities failed to take into account the peculiarities of the time. The enforcement of full-time work schedules put what had already been constructed into disarray. By simultaneously holding research and teaching jobs in different institutions, it was possible for Brazil's small scientific community to maximize productivity. A network bringing together scientists from institutes, colleges, public offices, and museums had been formed, often making it possible to overcome the material and technological limitations of each.

Cut off from the academic environment, and subject to the formal regulations and decreasing salaries of the public service, most institutions of applied research entered a period of decline. Scientific and university activities were doubly hit by the centralizing drive. First, they were victims of the attempt to unite and control the cultural and teaching scenes as a whole, carried on by the Ministry of Education. Then, as of 1937, they became victims of administrative unification, sponsored by the Departamento de Administração do Serviço Público, which believed that the scientific and educational system was simply part of a larger administrative body. "Desacumulação" made it clear that scientific activity had by itself not attained enough of its own distinct personality or autonomy for the powers-that-be to recognize a need to grant it special treatment or to recognize it as something valuable that needed to be

31. " 'Desacumulação' was decreed toward the end of 1937, beginning 1938.... [Formerly] a teacher or any other public servant could hold several posts simultaneously.... Of course this resulted in exaggerations that drew heavy public criticism. When the Estado Novo came to power, the so-called accumulation of posts was prohibited. Every employee had to opt for one definite post. This was done, I presume, with the best of intentions. I think each employee really should have only one job. But I believe 'desacumulação' was disastrous for Brazil in certain cases. At the same time 'desacumulação' was enacted, something should have been done to improve substantially everyone's professional situation. Posts were often accumulated not because employees wanted to have several different bosses but simply because they needed to earn more" (Leinz interview).

protected from political and bureaucratic vicissitudes. It was believed that whoever did scientific work at a government research institute, whoever taught at a public university, was first and foremost a civil servant, not a researcher or scientist. The aftermath of "desacumulação" made it clear just how fragile science was and just how little those enforcing the centralized, bureaucratic norms of the federal administration were aware of its worth and its special character.

Only those who somehow managed to escape from this general rule were able to succeed. Two striking cases, each reflecting a different way of dealing with the same situation, confirm this. The first case was São Paulo's Faculdade de Filosofia, Ciências e Letras, backed by the state of São Paulo's broader movement for regional autonomy. The second was the biophysics laboratory at the Faculdade de Medicina in Rio de Janeiro, which later became the Instituto de Biofísica, where Carlos Chagas Filho eventually managed to create research conditions for his handpicked group of aides. In both cases, it is clear that circumstances made it impossible to overcome difficulties and obstacles through the prestige and recognition of their scientific and academic worth; it was necessary to go the way of politics or to use personal and family networks to get through bureaucratic red tape and the lack of broader support. For Chagas Filho an aristocratic family background and connections to top-level administrative posts—hardly the norm among scientists—turned out to be crucial.[32] Scientific and academic work was therefore still unusual, and the exception rather than the rule, as Brazilian society moved into modernity.

32. "I was invited to join Manguinhos in the area of endemic diseases. This made it easy for me to meet with a minister—a very important minister of culture, Minister Capanema—and especially with one of the most dedicated public spirits I know, Luís Simões Lopes, director of the Departamento de Administração do Serviço Público, which was more powerful than it is today. It was Luís Simões Lopes who made it possible for me to hire Herta Meyer, Veiga Sales de Moura Gonçalves, and others by creating a special category of staff members designated specialized technicians. These employees could work thirty-three hours a week and earned more than a tenured professor—not a great deal more, but more" (Chagas Filho interview).

8

POSTWAR MODERNIZATION

Changes in Brazilian society since World War II can be described as a frenzied and often awkward march toward a future—the year 2000 is often mentioned—when the country was expected to enter at last the ranks of the modern, civilized, and rich nations. As this magic landmark gets closer it becomes obvious that, if there is such an opportunity, signs of crisis also abound and this chance may be lost.

Events in science, technology, and education should be seen in that light, as well as in light of the transformations that swept Brazilian society in the last decades, concentrating a rapidly expanding population in large urban centers, raising the general level of education, and replacing agriculture with an extended industrial economy (Table 9).

This period can be divided into two very different parts by the year 1968, when new graduate programs were created, undergraduate enrollments began to expand at very high rates, and much more money was allocated to research. The 1980s mark the beginning of a third and

TABLE 9. Structural Changes in Brazilian Society, 1950–1980

	Around 1950	Around 1980
Population in cities of more than 20,000 inhabitants	21.0%	46.0%
Employment in the primary sector	60.0%	30.0%
Technical, administrative, and similar occupations	10.0%	19.0%
Occupations in industries	13.0%	21.0%
Coffee as percentage of total exports	60.0%	13.0%
Industrialized products as percentage of total exports	—	57.0%
Literate population (10 years and more)	43.0%	74.5%
Population with 8 or more years of education (19 years and more)	1.9% (1940)	22.8%
Enrollment in higher education institutions as % of age cohort	0.9%	10.0%

SOURCE: Faria 1986:78; Castro 1986b:106; and Brazilian censuses.

different period, characterized by stagnation, crisis, and stock-taking regarding the achievements of the previous years.

In 1964 a politically conservative military government came to power, and its relationship to the scientific community and the universities tended to be highly conflictive, culminating in the early 1970s with the dismissal of hundreds of scientists and professors from their jobs and exile for many. The predictions that the new regime would be completely closed to new ideas on science and education, however, did not materialize. In 1968 higher education went through deep reorganization and entered a decade of rapid expansion. Also in the late 1960s new agencies were created, and funds for science and technology began to be organized, leading to an unprecedented expansion of graduate education and research institutions. The contradiction between these policies and the simultaneous repressive measures reflect, in good measure, the lack of any clear policy on scientific and educational matters coming from the central government, leading to decisions based on a division of spheres of influence within the state's bureaucracy. Political repression was at its highest when a military junta assumed full powers in late 1968 and for several years gave free rein to the so-called intelligence and repressive groups within the military; it became more limited during Ernesto

Geisel's presidency after 1975, when an ambitious project of national growth was attempted, and the more repressive groups were curtailed in their action.

Brazil's military participation in World War II was not very extensive, but it provided an opportunity to attempt the first program of economic mobilization and planning in the country's history. Traditional patterns of trade were upset, but Brazil became an important supplier of some important strategic materials for the Western Allies—diamonds, manganese, nickel, tungsten, and more important, rubber. To ensure a supply of these products, American assistance was provided to equip laboratories and organize production. An important step in the country's industrialization was the creation of the steel mill of Volta Redonda, with technical and economic support from the United States, as part of the agreements that brought Brazil into the war.[1] The reduction of imports increased the demand for São Paulo's manufactured products and created a surplus of foreign currency. When the war ended, a constitutional government based on universal suffrage replaced the Vargas regime, and Brazil's economic surplus helped create a market for industrialized goods. Once the foreign reserves were depleted, Brazilian and foreign-owned industries began to produce locally for this market in an expanding urban society.

Scientists as Intelligentsia

Optimism about the positive role science and technology could play in bringing the Latin American countries to higher socioeconomic levels was high in the first years after World War II. The war had shown the power of science and technology for destruction and led to the hope that they could have an equally strong impact if properly oriented. The wave of technological changes in industry and agriculture only seemed to confirm this idea.

The view that science and the universities could have a positive role in socioeconomic change was part of the "developmentalist" ideology that emanated from the United Nations Commission for Latin America.[2] In a document published in 1970, Raúl Prebisch emphasized the need to adapt and regroup international technological knowledge to meet Latin America's specific conditions. He believed in establishing priorities from

---

1. McCann 1973.
2. For a full discussion, see Schwartzman 1984b and 1985.

an economic planning point of view and in organizing research programs to respond to those priorities. "All this has a close relationship with education. It will be necessary to promote educational programs that, besides the diffusion of technologies, should have as one of their main purposes stimulation of the creative capacity in this field."[3]

Postwar scientistic activism was different from that which prevailed during the organization of the Faculdade de Filosofia at the Universidade de São Paulo. Before the war the need for science was proclaimed in the name of culture, civilization, and leadership. Later, science began to be perceived as an important tool for economic development and planning, and scientists argued that they had a responsibility to avoid limiting themselves to the academic life. They wanted to participate in all relevant decisions in their societies, and they felt capable of doing so. The involvement of scientists in England, the United States, and the Soviet Union in the war effort had been followed closely, and the ideas put forward in previous years by J. D. Bernal and Frédéric Joliot-Curie were well known. The experiences with military research at the Faculdade de Filosofia during the war also helped.

The proponents of this new scientific role were highly qualified people, usually with work and study experience in Europe or the United States. They had encountered other cultures and mentalities and did not accept the prestige hierarchies of their societies; they were confident about their ability to change and lead a modernized educational and research system, given enough international and national support to try out their ideas. They believed that a scientific approach should be put to work not only for developing new technologies or controlling tropical diseases, but also for implementing social and political planning at the highest possible level. Political participation was generally perceived as a necessary channel for reaching the levels of influence and social responsibility the scientists thought they needed. Their political outlook tended to be rationalistic, nationalistic, and socialist.

Several lines of action followed from these premises: the scientific community should be organized and mobilized; the educational system should be changed; science and technology should be provided with strong, institutionalized planning agencies; and specific policies for science and technology should be put forward with all the political support that could be mustered.

The first step in the organization and mobilization of scientists was the establishment, in 1948, of the Sociedade Brasileira Para o Progresso da Ciência (SBPC), the Brazilian counterpart of the associations for the advancement of science that existed in different countries. The main

---

3. Quoted in Graciarena 1964.

organizers were Jorge Americano, José Reis, Paulo Sawaya, Maurício Rocha e Silva, José Ribeiro do Vale, and Gastão Rosenfeld, all from São Paulo's biological research institutions. Its first, short-range purpose was organizing São Paulo's scientific community in defense against the populist politics of the governor of the state, Ademar de Barros. Later, it developed a series of activities aimed at strengthening its role as the national representative body of Brazilian scientists: promotion of annual meetings in different Brazilian cities; publication of a journal, *Ciência e Cultura,* for distribution among its associates; support for the creation of specialized scientific associations, which would usually hold their annual meetings jointly with the SBPC; and a pattern of close contacts and association with the international scientific community, on the one hand, and with Brazilian science policy and scientific authorities on the other.

For a few years in the 1970s, the annual meetings of the SBPC were the only open forum for discussion of all kinds of questions in an otherwise closed and strongly censored political regime. Because of that, the association gained public notoriety, and its meetings attracted thousands of participants and substantial press coverage. Meanwhile, strictly scientific questions tended to move toward the specialized scientific associations, leaving the SBPC mostly to its broader representation and intermediation roles. In the 1980s the SBPC began publishing *Ciência Hoje,* a new and highly successful magazine based in Rio de Janeiro and aimed at the dissemination of science and the work of Brazilian scientists among the educated public.[4]

Another important event was the institutionalization of the Fundação de Amparo à Pesquisa do Estado de São Paulo (FAPESP, São Paulo Foundation for Research Support), an agency established by the state's constitution of 1946 but organized only in the early 1960s. Endowed with 0.5 percent of the state's tax revenues, forced by statute to spend most of its money on actual scientific research projects, and directly controlled by the state's scientific community, the FAPESP became Brazil's main financing alternative to the federal agencies established in the 1950s and 1960s with similar purposes.

Nuclear Energy and the Conselho Nacional de Pesquisas

In 1949 a private research institution, the Centro Brasileiro de Pesquisas Físicas, was established in Rio de Janeiro. Intended to start Brazil along

---

4. M. R. Silva 1960 and 1978; Botelho 1983.

the road of atomic research without the constraints typical of educational institutions or the civil service, it brought together several high-quality scientists—Cesare Lattes, who returned to Brazil especially for this purpose, José Leite Lopes, Jaime Tiomno, and Roberto Salmeron. In 1951 a governmental body under direct supervision of the president of the Republic was created to support science: the Conselho Nacional de Pesquisas (CNPq, National Research Council). Both institutions were created by the personal efforts of Admiral Alvaro Alberto da Mota e Silva, a military man who looked at science and technology from a strategic point of view and a mathematician and physicist of some standing. A national commission on atomic energy, the Comissão de Energia Atômica, was supposed to be organized within the research council. The understanding was that the commission would set the policy guidelines and that the council would see that the research center had the resources it needed to carry out its assignments.

In 1953 the Instituto de Pesquisas Radioativas was organized in Minas Gerais, and from 1956 on it operated an experimental reactor of the Triga type using enriched uranium. In 1958 a group of physicists from this institute began work on a nuclear reactor based on thorium as fuel, but that project was discontinued when the government decided to acquire a Westinghouse plant based on enriched uranium. In Rio de Janeiro, besides the Centro Brasileiro de Pesquisas Físicas, the Universidade do Rio de Janeiro started its first course on nuclear engineering in 1954, which graduated fewer than one hundred persons in its first ten years. In 1965 it began to operate its own 10 kilowatt experimental reactor. In 1956 a much larger Instituto de Energia Atômica was organized within the Universidade de São Paulo, with a 10 megawatt swimming pool reactor and about one thousand research workers and technicians. In 1971 a 22 megawatt particle accelerator was installed.

Despite these promising omens and the quality of its work in other fields, the Centro Brasileiro de Pesquisas Físicas never really began to work in the field of atomic energy, and after its first few years it fell victim to serious institutional problems. The original efforts of the research council in the field of atomic energy were supported only briefly during the second government of Vargas, which ended with his suicide in 1954. The United States did not support these efforts because it wanted to retain control over the enrichment process, and in 1954 the U.S. government stopped the delivery of three centrifuges for uranium enrichment that Brazil had purchased from the University of Göttingen in West Germany. Finally, the limited scale of the research effort begun by Brazil in those years raises doubts about whether it could have accomplished much of significance.

Once established, but deprived of its main objective, the Conselho Nacional de Pesquisas became an agency that distributed its limited resources to individual scientists in the biological, physical, and other natural sciences. With that agency's backing, small-scale, independent research could develop in a few centers, even when the scientist's own university—usually more concerned with the problems of professional education or narrow, short-term technical investigations—offered little support. Besides supporting research, the agency provided, as it still does, travel grants and fellowships for postgraduate and advanced studies abroad in conjunction with the Coordenação de Aperfeiçoamento de Pessoal de Nível Superior (CAPES), an agency of the Ministry of Education. Endowed with only a small staff, the CNPq based its decisions on advice from the scientific community, which assured competent use of its limited resources. In the late 1970s the CNPq changed its name to Conselho Nacional de Desenvolvimento Científico e Tecnológico and came under the Ministry of Planning.[5]

The New Elite Universities

An important feature of the postwar period was the creation of a few elite teaching and research institutions, which affected only a small portion of the growing higher education system but served as models and inspiration for broader reforms that would be attempted later. Detailed examination of some of these experiences reveals common features. They all had well-defined personal leadership; their origins and inspiration can be traced to some of the most significant groups, traditions, or institutions of the 1930s; and they were able to protect themselves from the equalizing pressures emanating from the Ministry of Education. Finally, they were all created anew and did not have to contend with entrenched interests and institutional routines.

The first of these institutions was the Instituto Tecnológico de Aeronáutica (the Aeronautical Technological Institute, ITA), which was part of a broader technological center created by the Brazilian air force. The Centro Tecnológico da Aeronáutica was conceived from the beginning as a military engineering institution, meant to provide technical and professional support to the Brazilian air force, which was created as an independent service in 1941. The project, approved by the Brazilian government in 1945, included an engineering school (the Instituto Tecnológico de

5. Romani 1982; Albagli 1987.

Aeronáutica) and a research center (the Instituto de Pesquisas e Desenvolvimento). American institutions were used as models, and the center's official history cited as appropriate models such places as the Massachusetts Institute of Technology (M.I.T.), the California Institute of Technology, and different research establishments belonging to the American air force, navy, and civil aeronautics establishment.[6]

The institute was organized in close cooperation with M.I.T., which sent one of its professors, Richard H. Smith, to coordinate the project. In the 1950s the institute became known as Brazil's best engineering school; it drew students from all over the country through very competitive entrance examinations. Because of its location—São José dos Campos, near São Paulo—the students had to live on campus, something that occurred at only a few agricultural schools. The institute was not organized as a military establishment and was open to civilians. Its location under the Ministry of Aeronautics freed it from the bureaucratic regulations of the Ministry of Education and provided it with many more resources than any other teaching institution in the country. The close cooperation with M.I.T. assured a constant flow of personnel between the institute and several American institutions and made it easy for its best students to continue their courses in the United States.

The physics department of the new institute was headed by Paulus A. Pompéia, a former assistant to Wataghin. Pompéia recalls the names of the people involved in the project—Ernesto Luís de Oliveira Jr., who had worked with Luigi Fantappié; Air Force Colonel Casimiro Montenegro Filho, the entrepreneur behind the project; and Richard Smith, the first reactor. The novelties of the project enticed Pompéia to come to the institute: full-time teachers and students, prospects for long-term careers for professors, and resources for research. At first most of the professors came from the United States. After the first ten years the physics department had about fifty professors, and there was a strong emphasis on experimental research.

The core group—Walter Baltensberger of Swiss origin, Sérgio Porto, Luís Valente Boff, Mário Alves Guimarães, and José Israel Vargas—began a new tradition of solid-state physics that had existed previously in Brazil only through the work of Bernhard Gross. It was the beginning of a new generation, and the list of former students includes José Ellis Ripper, Rogério Cerqueira Leite, Heitor Gurgulino de Souza, João Bosco de Siqueira, Geraldo Aurélio Tupinambá, and Anísio dos Santos.

6. Paim 1987:13–14.

For Sérgio Porto this was a radical and badly needed departure from the tradition of particle physics inaugurated by Wataghin.[7]

The new institution was not easily accepted, but its military support helped. Pompéia recalls:

> The first problem we had was with the Ministry of Education, which did not understand that an engineering school could be outside its supervision. I was in charge of the negotiations with the ministry.... A primary goal for the institute was to create an engineering school patterned on the American schools. That meant to mold practical engineers, not theoreticians. The Politécnica had a very strong French influence, and the Polytechnique in France was more a science school than an engineering school.

Since there was never an agreement, the diplomas provided by the institute were registered only by the Ministry of Aeronautics, which in the end did not cause any difficulties for its students.

Resistance came also from the military brass, which did not accept easily the idea that its school should produce civilian engineers. The original project was to have had a purely military establishment.

> Richard Smith sent a memo to Brigadier Montenegro and [Aeronautics] Minister Trompowsky showing that this would be a waste of resources, that the Ministry of Aeronautics was responsible for the development of Brazil's industry; and that they needed civilians because they could not do it only with military officers. Since they were spending so much to build the school, they should have had 90 percent civilians and 10 percent military personnel among the students, with the advantage that these military men, who would hold key positions in the future, would get the chance to know the civilians with whom they studied.[8]

The prevalence of this conception helps explain the contrast between the ITA and its army counterpart, the Instituto Militar de Engenharia in

---

7. "A society cannot have only poets. It needs people concerned with its national needs.... I trained only solid-state physicists, spectroscopy people.... It has been a long battle, but you can see that today solid-state physics, my physics, dominates the whole country" (Porto interview)

8. Pompéia interview.

Rio de Janeiro, which remained a purely military establishment. According to Pompéia, however, most air force authorities were against this conception of the school. As early as 1960 there was an attempt to organize the institute as an independent foundation, a project inspired in what was being proposed for the Universidade de Brasília. The effort was barred by the military and led to Pompéia's decision to leave the institution.[9] The tendency toward militarization became irresistible after 1964 and led to the resignation of Pompéia, who moved to São Paulo's Instituto de Pesquisas Tecnológicas.

From the beginning the engineering school was part of a larger technological center whose institute for research and development was headed by Brigadier Aldo Vieira da Rosa. In 1971 the Centro Tecnológico changed its name to Centro Técnico Aeroespacial (CTA), and technological research increased its importance. In the mid-1980s the center had about 5,400 employees, 1,100 with university degrees. Its research activities included rocketry and artificial satellites. Besides its own teaching and research activities, the CTA provided the conditions for the creation of EMBRAER, Brazil's state-owned airplane manufacturer. Several high-technology institutions, public and private, were established in its surroundings, forming Brazil's closest approximation to a "Silicon Valley" phenomenon. The engineering school itself, after a severe crisis and the loss of many of its civilian staff, lost some of the luster it had in the 1950s and 1960s.

A second experience involved the Faculdade de Medicina de Ribeirão Preto, founded and led by Zeferino Vaz in a small city at the heart of São Paulo's coffee country,[10] which was to become one of Brazil's best medical schools. Demand for medical education was on the rise, and the traditional Faculdade de Medicina of the Universidade de São Paulo did not want to expand. Vaz was the director of the new institution until 1964, when he moved to the Universidade de Brasília.

Vaz explains his success by what he calls "an open secret":

> I know how to attract a scientist. I learned it early while working in Travassos' laboratory and at the Instituto Biológico under terrible conditions. Scientists are attracted when you offer them a new scientific ideal. What I offered them was a complete revolution in

9. Porto interview.
10. Zeferino Vaz, born in 1908, had a medical degree from São Paulo's Faculdade de Medicina and studied parasitology with Lauro Travassos, general biology and genetics with André Dreyfus, and zoology with Hermann von Ihering. He worked as a researcher at the Instituto Biológico between 1929 and 1937 and was professor of parasitology at the Universidade de São Paulo after 1935.

medical education. Why revolution? Because in those years the physicists had created very sophisticated instruments for analyzing biological phenomena. . . . However, this revolution had not been incorporated into the teaching of medicine, which remained mostly morphological and static, based on a three-year anatomy course. There was the cult of the cadaver.

The new emphasis on biochemistry, physiology, and pharmacology; the organization of disciplines into academic departments; and active recruitment of talented people were the ingredients Vaz used in his project. New disciplines were introduced into the curriculum; others lost their relevance. Anatomy was reduced to one year, while pediatrics, gynecology, and obstetrics increased their load. Preventive medicine and medical psychology were introduced.

I also introduced biostatistics in the medical course. Why? Because this basic contribution of physics made it possible to quantify the biological phenomena. . . . To study the variations of normality and disease under different conditions—this is what I call the Galilean era in biological sciences, through mathematicization. Thus, the biological sciences, which were mostly descriptive, are becoming like the exact sciences.[11]

To carry on his project, Zeferino Vaz had to confront the authorities at the Ministry of Education. "I took the new plan to Jurandir Lodi, who was the dictator of higher education. 'Oh, you cannot do that.' 'Why not?' 'Because you have to follow the model of the Faculdade Nacional in Rio de Janeiro.' 'Why do I have to follow a model that has been obsolete for fifty years?' 'Because it is written in the statutes that this is the model school for medical education in Brazil.' "[12] It was necessary to lobby the members of the Conselho Federal de Educação to have the projects approved. That the new school did not depend on federal support was decisive.

Thanks to his personal prestige, the working conditions offered, and the new professional and research perspectives opened, Vaz was able to attract a group of outstanding professors from Brazil and abroad.[13]

---

11. Vaz interview.
12. Vaz interview.
13. Including J. Moura Gonçalves, Maurício Rocha e Silva, Lucien Lison, Miguel Covian, Fritz Köberle, J. L. Pedreira de Freitas, Mauro Pereira Barreto, and J. Oliveira Almeida.

Support came not only from the state budget but also from the Rockefeller Foundation and other sources. Well conceived, properly endowed, and limited in its ambitions, the Faculdade de Medicina de Ribeirão Preto retained its quality and prestige after Zeferino Vaz departed and its endurance and stability make a positive counterpoint to the usually short productive cycle of Brazilian academic institutions.

The third experience involved the Universidade de Brasília. This university was part of the whole project for the country's new capital, and its organization was entrusted to Darcy Ribeiro.[14] The first study group designated to organize the new university was formed by Ribeiro; his close friend and Juscelino Kubitschek's chief of staff, Ciro dos Anjos, a writer; and architect Oscar Niemeyer. Anísio Teixeira also participated in the project from the beginning.

Ambitions were extremely high, and for Darcy Ribeiro the past, in Brazil as elsewhere, was *tabula rasa*. He describes the new university as

> the most ambitious project of the Brazilian intellectuals—a project to revise the culture of the world, knowledge, science, and scholarship and to discover what scholarship, knowledge, and science could give us. Brasília was a radical attempt to rethink the university all over again, that old, archaic, millenary sacred cow. . . . I only asked people who were dissatisfied to work on the project, those who wanted the university to be what it should be, not those who wanted to reproduce what it was, here or anywhere else in the world.[15]

This new university was to have different roles. First, it should provide cultural substance to Brasília, a city built in the middle of nowhere; second, the university should become a "superadvisory agency to the government, without being subservient, without being a group of government employees, but retaining its autonomy as a cultural institution. . . . It should be the big advisory agency." Third, it should provide Brasília with its spirit, its creativity. Last,

---

14. Ribeiro was an anthropologist from Minas Gerais who studied at the Escola de Sociologia e Política de São Paulo with Emílio Wilhems and Herbert Baldus. In the 1950s he became associated with Anísio Teixeira at the Centro Brasileiro de Pesquisas Educacionais (Brazilian Center for Research on Education) in Rio de Janeiro. In the 1960s he was President João Goulart's chief of staff, and in the 1980s, after several years of political exile, he became vice-governor of Rio de Janeiro under Leonel Brizola, as well as Brizola's candidate for the 1986 gubernatorial election in that state.

15. Ribeiro interview.

> This university should give Brazil its first opportunity to reach excellence in all fields of knowledge. . . . All fields of knowledge had to be cultivated and cross-fertilized. If we could have good chemistry side by side with good mathematics and physics, it would be possible to produce people who could use the scientific way of thinking to deal with the country's problems, rather than an ancillary thinking typical of those second-rank assistants we used to prepare in the country.[16]

The treatment had to be radical:

> An important science policy decision was to forbid the Ford Foundation to do what it used to do with Brazilian science. The foundation would give some money to each Brazilian scientist to hire some assistants. . . . It would give him some money to buy equipment or to supplement his salary. He would become an appendix, because the scientist would be linked to a foreign professor, usually American (but he could also be English) who would come here occasionally. . . . It is not that the foundation wanted to colonize Brazil this way; it believed this was the best way to help. . . . In Brasília, from the beginning, we forbade the Ford Foundation or any other organization to deal directly with the professors. Any financial support should be dealt with by the rector's office; we would not allow the entrepreneurial professor to look for his money here and there, which is something that deforms the institutions. But I had very important support from the Ford Foundation—more than $2 million to buy a basic sciences library of more than 150,000 volumes.[17]

The university was organized around a series of central institutes divided along disciplinary lines, each responsible for undergraduate and graduate teaching and research. The chair system was not to be adopted, and the institutes were to have a collegial organization. Formal power, however, was concentrated at the top. From a legal point of view, the university was established as an autonomous foundation and granted a large endowment of real estate and shares of publicly owned companies.

There was no chance to see how these ideas would work in practice. At

---

16. "Ancillary science" was a scourge to be avoided at all costs. In the old universities in Brazil, as well as in other underdeveloped countries, "you could have very good biochemistry, but it was linked to a given group in Germany or England. It was an appendage, a slave working here on problems decided outside. It was a crazy biochemistry" (Ribeiro interview).

17. Ribeiro interview.

first Brasília attracted many young professors and scientists who were, as Ribeiro indicated, dissatisfied with the Brazilian academic institutions. For a while the entire group of physicists from the ITA considered transferring to the new university. Darcy Ribeiro himself left the Universidade de Brasília in 1962 to join João Goulart's government as minister of education and later as the president's chief of staff. He was replaced by Anísio Teixeira, who remained until the military coup of 1964, when he was replaced by Zeferino Vaz.

In 1964 Zeferino Vaz had left Ribeirão Preto to become São Paulo's state secretary of health "in the preparatory stages of the 1964 revolution." His conservative outlook, combined with his academic credentials, made him a rare asset for the Brazilian military regime, and in April of that year President Castelo Branco asked Vaz to become rector of the Universidade de Brasília. Vaz describes this experience as a war on two fronts: in defense of quality and against external intervention. He acknowledges that he dismissed "seventeen or eighteen elements"—most of them social scientists recruited by Darcy Ribeiro at the Universidade de Minas Gerais—not because they were Communists but "for mediocrity." He claims to have resisted external pressures against competent people and to have supported the work of such people as musician Cláudio Santoro, architect Oscar Niemeyer, Vice-Rector Almir de Castro, and mathematician Elon Lages de Lima. Newly invited professors included Roberto Salmeron in physics, Otto Gottlieb in chemistry, and Antônio Cordeiro in biology.

In 1965 Zeferino Vaz left Brasília to work in the organization of the Universidade de Campinas, probably anticipating the storm that was to come. In spite of his intent to keep the Universidade de Brasília free from ideological confrontations and external interference, a series of dismissals and resignations emptied the university of more than two hundred of its professors. More than that, the university lost its credibility among the Brazilian academic community. Although it remained among the best in the network of federal institutions—thanks to its physical installations, innovative organization, financial endowments, and the quality of some of the remaining staff—it would never regain its initial mystique and prestige.

Expansion of Higher Education

The idea that the new elite universities were forerunners of a deep transformation of higher education captured many minds. This proposal required, however, a dramatic change in most of the country's existing higher education institutions. It went against the gradually growing trend toward mass education, and it required breaking the power of old faculties, imposing demanding patterns of scholarship on students and teachers, placing more value on research work than on professional achievement, and discriminating within higher education between good and bad universities, departments, research groups, and courses. It also meant dividing the students into those who would be oriented toward research and those who would be limited to conventional education for the liberal professions.

Rapid urbanization, mass communications, and mass consumption, however, were leading the universities in a different direction. People wanted more education and privileges associated with it, but not necessarily more demanding courses. There was, of course, effective demand for more engineers, lawyers, doctors, and teachers. Less recognized but probably more important was the desire of the middle classes for social prestige and the benefits professional status brought. A university degree promised a certain level of social prestige and income regardless of the quality of education received. In time, legal privileges for diploma holders were established not only for the traditional professions—physicians, lawyers, and engineers—but also for new professions like economists, statisticians, administrators, journalists, librarians, and psychologists. To respond to the demands, the federal government built a network of federal universities that often absorbed old state and local institutions that could not sustain themselves and expand with their original resources. With some minor exceptions, only the state of São Paulo kept its own institutions of higher education. Private institutions also emerged—first the Catholic universities organized by the church, and afterward a large variety of religious, lay, community, municipal, and privately owned institutions, all under the nominal supervision of the Ministry of Education and its Conselho Federal de Educação. Universities were supposed to be autonomous, while isolated establishments were supposed to come under federal supervision. However, the universities were bound to curricula established by legislation for their professional degrees, and the federal establishments' budgets were strictly controlled by the Ministry of Education while their professors came under the civil service statutes. The chair system guaranteed that professors could not be fired and could teach freely without interference, and in each school a faculty council formed by chair holders had the final say on all matters not conflicting with federal rules and regulations. The faculty councils also drew up the list of names from which the government appointed the schools' directors, while university-wide councils drew up the list of names from which the government would appoint the rectors. In such a system, most of the power remained with the schools; the rectors had mostly a ceremonial role.[18]

This system of higher education enrolled in 1968 about 278,000 students, less than 5 percent of the nation's twenty- to twenty-four-year-olds. (The total Brazilian population for that year was estimated at 87 million.) Enrollment at the high school level was around 800,000, and at the primary level (until the eighth year) there were about 14 million students, mostly concentrated in the first four years of education. Fifty-five percent of the students were in public, tuition-free institutions, most of them belonging to a university; the remaining 45 percent were in private establishments, most of them isolated schools without university status. (The degrees provided by universities or isolated schools, however, are equivalent, and both are considered "university" degrees. The only differences are institutional: universities are supposed to be freer from ministerial supervision and can have larger bureaucracies.) In terms of subjects of study, about 25 percent were in "soft" fields, such as the humanities, literature, and the social sciences (mostly in the schools of philosophy, sciences, and letters); about 20 percent were in law; 10 percent were in medicine; and another 10 percent in engineering. Admission for higher education came through public examinations given by each institution and open to high-school-degree holders. There were 2.4 applications for each place in 1968, with much higher ratios for the established professions in public universities.

From this basis an ambitious project to eliminate stages and bring Brazil directly into the twenty-first century was attempted. In the last two chapters, we examine this "great leap forward" and its aftermath.

18. For an expanded discussion, see Schwartzman 1988a. See Levy 1986 for a comparative view of higher education in Brazil and other Latin American countries.

9

THE GREAT LEAP FORWARD

Science and Technology for Economic Development

The involvement of Brazil's main investment bank—the government-owned Branco Nacional de Desenvolvimento Econômico—in the field of science and technology is the most important feature of the new period. For the first time in Brazil's history there was a concerted attempt to put science and technology at the service of economic development, through the investment of substantial resources. In 1964 the bank established a program for technological development known as the Fundo Nacional de Tecnologia (National Fund for Technology), which in its first ten years provided about $100 million for research and graduate training in engineering, the hard sciences, and related fields.

The Fundo Nacional was established with the hope that economic incentives would lead private investors to develop their technologies instead of importing them from abroad, and it soon began to support selected teaching and research programs. With the fund's support, the

Universidade de São Paulo acquired its electrostatic accelerator, Pelletron, in 1971; a consortium of institutions began to develop a Brazilian minicomputer; the Centro Tecnológico da Aeronáutica obtained support for its work on airplane engines; the Instituto Militar de Engenharia initiated graduate programs in several branches of engineering and chemistry; and the newly created Universidade de Campinas received substantial grants for a variety of projects. The creation of a complex system of graduate courses in engineering at the Universidade Federal do Rio de Janeiro, which came to be known by the acronym COPPE, was an important initiative. The activities of the fund were later transferred to a new, specialized agency, the Financiadora de Estudos e Projetos, which works as an investment bank for technological and feasibility studies and which administers a national fund for science and technology that became part of the federal budget, replacing the development bank's fund. In 1975 the old Conselho Nacional de Pesquisas was transformed into a new and much larger Conselho Nacional de Desenvolvimento Científico e Technológico, now under the Ministry of Planning.

The ideological roots of this program can be traced to a combination of two seemingly opposing trends. First, there were the ideas about economic and technological dependency and the need for science planning that were so central to the Brazilian scientists' quest for a new role. José Pelúcio Ferreira—the economist who organized the Fundo de Desenvolvimento, FINEP, and later became vice-president of the national research council—acknowledges the role of physicist José Leite Lopes in shaping his views. Adler reported:

> [Pelúcio Ferreira] said that by the middle of the 1960s, although physicists and economists developed their ideas separately they had converged to create an awareness of science and technology dependency. The economists' emphasis on the linkages between technology and economic development was particularly important. Pelúcio has acknowledged that both *idéias cepalinas* and the ISEB [Instituto Superior de Estudos Brasileiros, a think tank in Rio de Janeiro closed by the military in 1964] had a considerable effect on his subsequent work in the science and technology field.[1]

The other trend was the military government's nationalist ambitions, which began to take shape after the 1960s and peaked in the mid-1970s. In the 1960s, South American military regimes were known for their

1. Adler 1987:210; Lopes 1978.

ideological and doctrinaire approximation of the United States, for their economic liberalism, and supposedly for their concern with reducing the role of the state in all spheres of activity—except of course for control of political participation and expression. In that sense, they were opposed to the trend represented by CEPAL or such scientists as Leite Lopes, who supported increasing state planning and intervention to redress the effects of dependency. The most extreme example is probably Chile, which became a test case for Chicago-style economics orthodoxy. Economic liberalism was also central to the first Brazilian military regime headed by General Castelo Branco after 1964. The orthodox economic policies of those years were effective in controlling inflation, increasing the government's tax base, modernizing the government's instruments of economic policy-making, and attracting foreign capital.

Economic liberalism was followed shortly by a parallel, and eventually opposite, tendency toward growth and strengthening of the public sector. In the 1970s a division of labor began to shape up. The liberal economists would continue to run Brazil's economy; political participation would remain under control; and the military, the engineers, and eventually the scientists would speed up their long-term projects through the expansion of the state. The list of projects was impressive: the nuclear program, the large hydroelectric power plants, several ambitious road and railroad construction projects, the expansion of the frontier into the Amazon. The consequences, both positive and negative, are still being evaluated. On the positive side were the modernization of the country's industrial base and the effective growth of national income; on the negative side were the excessive levels of income concentration, the destruction of the environment, the emptying of the countryside, the deterioration of cities, the swelling of the state, the wastefulness of unfinished and overambitious projects, and the economic indebtedness that led to the economic crisis of the 1980s.[2]

What brought intellectuals and scientists on the left together with the military on the right was nationalism and the shared belief in the powers of science and technology. It was not an easy coexistence. Many intellectuals lost their academic positions and were forced into exile. The building of new scientific and research institutions, and the participation of talented scientists who did not submit easily to military authoritarianism, required constant and difficult negotiations with security officers, which were conducted, not always successfully, under the authority of the minister of planning, João Paulo dos Reis Velloso, or by people like Zeferino

2. Lessa 1978; F. M. de O. Castro 1985; Schwartzman 1980.

Vaz. There was also a clear contradiction between the economic policies being pursued by the Ministry of Finance, oriented toward the internationalization of the economy and the introduction of foreign capital and technology, and the projects for technological self-reliance carried on under the Ministry of Planning. Because of this split, investments in technology were seldom based on broader macroeconomic considerations, while economic policies never took the development of national technologies into account.

The entrance of agencies of economic development and planning into the field of science and graduate education intensified the historical tendency to favor applied technology over basic science, a tendency that was dramatized by the change in the name of the Conselho Nacional de Pesquisas, which was far from just nominal. Science-supporting agencies like FINEP and the science council gradually became swollen bureaucracies of hundreds and eventually thousands of functionaries, and scientists had to negotiate with economists and administrators every two or three years, on a project-by-project basis, for renewal of their grants. A two-year national plan for science and technology was promulgated in 1973 and again in 1975, with projected expenditures ranging from $323 million to $824 million a year.[3] These plans were little more than collections of anticipated expenses by sector, most of which—65 percent for the period 1973–75—were completely outside the sphere of influence of the planning authorities in such agencies as FINEP, the national research council, or the Banco Nacional de Desenvolvimento. The expectation for the period 1976–77 was that the expenditures under these agencies would be increased slightly. Between 21 percent and 27 percent of the expenditures were to be allocated to graduate training, fellowships, and "scientific development" in general; between 20 percent and 29 percent to industrial technology; between 11 percent and 15 percent to agricultural research; and between 5 percent and 10 percent to atomic energy projects. There is no known evaluation of how the plan was implemented or how the expenditures were made. The third plan, for the years 1978–79 and already under President João Batista de Figueiredo, was just a broad statement of purposes without any figures attached. At that time João Velloso and José Pelúcio Ferreira had already left their posts, and Delfim Neto, the former finance minister, was running the economy from the Ministry of Planning. For the first time, macroeconomic and technological policies came under the same authority, which gave the latter very low priority.

3. Schwartzman 1978:574.

The 1968 Reform of Higher Education

In 1968 new legislation aimed at a profound reorganization of higher education was introduced. The number of applications was increasing, and it was impossible to keep the system small. This was also a time of intense street demonstrations against the military government, which ushered in several years of student-based urban guerrilla activity and violent government repression, including tight control over political activities at the universities. The 1968–78 decade was also a period of rapid economic growth, with new jobs being created and social mobility intensifying. When combined, these factors led to a complete revamping of the country's higher education, although not necessarily in the directions prescribed by the 1968 legislation.

The 1968 reform adopted the ideas developed in the mid-1960s by the Universidade de Minas Gerais—which were in turn based on the experience of the Universidade de Brasília in the early 1960s—and responded to the desire of highly educated groups to adopt the American research university model.[4] It is possible to establish a direct link of ascendance between these ideas and those tried in the frustrated experience of the Universidade do Distrito Federal: Anísio Teixeira was part of both, and Darcy Ribeiro, who organized the Universidade de Brasília, was associated with the Centro Brasileiro de Pesquisas Educacionais during the 1950s.[5] There was also a more direct American presence, through the advice of a joint commission established between the U.S. Agency for International Development and the Brazilian Ministry of Education.[6] From an organizational point of view, the 1968 reform introduced many elements taken from the North American research universities: the departments, which led to the elimination of the traditional chair system; the credit system, which did away with serialized, year-by-year course programs; research institutes; graduate programs providing master's and Ph.D. degrees; and a "basic cycle" in the universities, which was designed to provide some kind of general, college-like education in the first two years of school. All higher education institutions were supposed to evolve to this model; no room was allowed for institutional and role differentiation.

Implementation led to unanticipated results, partly because the 1968 reformers did not foresee the explosion in demands for higher education that was gaining speed precisely in those years (see Table 10). Appli-

 4. Pimenta 1984:24.
 5. Mariani 1982a.
 6. Carneiro et al. 1969.

TABLE 10. Growth of the Educational System in Brazil, 1965–1980 (1970 = 100)

	1965	1970	1975	1980
Population	87.3	100	115.0	127.8
High-school graduates	49.3	100	163.1	239.6
Vacancies for higher education	39.7	100	240.1	279.3
Applicants for higher education institutions	33.7	100	237.5	548.5
Undergraduate enrollment	34.2	100	212.0	294.9
Enrollment in private institutions	28.3	100	245.3	353.7
Enrollment in public universities		100	164.0	238.2

SOURCE: Education data from Ministério da Educação, Serviço de Estatística da Educação e Cultura; population data from Brazilian censuses.

cations for higher education institutions increased more than fivefold between 1970 and 1980, partly because of the expansion in secondary education and partly because new social groups (women, older people) were trying to enter the system. Government authorities responded to the pressure by letting private institutions of higher education proliferate without much quality control, moving further and further away from the research university model that was supposed to be followed by all.[7]

The new rules were applied mostly to the public sector, but even there the results were different from what was expected. Previously, power resided mostly with the schools' congregations, and the old professional schools were the only real institutions from a sociological standpoint, within universities or in isolation. The new arrangement sought to transfer power to departments and research institutes and to transform the old professional courses into simply a sum of credits to be obtained by the students in different departments. The stronger and more traditional schools resisted this change and implemented the department-institute organization only within their walls. New and weaker areas were more open to innovation, but their very weakness led to power concentration in the rectors' offices. The schools that kept their institutional integrity were the ones that best managed to maintain or improve their quality in the years to follow.

Two other innovations—the basic cycle and the credit system—also had dubious results. Students finishing the equivalent of high school had

7. Schwartzman 1988a.

to take competitive examinations for the career and school of their choice, and those who were admitted were immediately committed to those careers. The basic cycle became sandwiched between the entrance examinations and the vocational courses and was perceived by most as just an annoying waste of time. The credit system ran against the fixed and regulated contents of most careers and collided with the limited resources the institutions had for offering course choices. Both the credit system and the basic courses became, at best, new and more complex ways of doing the same old things and, at worst, administrative and pedagogic nightmares.

The educational authorities also created legal and budgetary conditions that allowed the universities to hire full-time faculty. In the past, university salaries had been low and had not competed with what a successful liberal professional could get from clients. As the universities expanded, new and nontraditional courses were introduced, graduate and research programs were created, and a new stratum of full-time faculty started to emerge. In part, members were called to staff the new graduate programs, but not many had the appropriate qualifications to work at this level, and the new basic cycle, coupled with the expansion of enrollments, required that a large number of new teachers be admitted without delay. The result was that, within a few years, most of the faculty of the Brazilian public universities went from part-time to full-time without necessarily increasing its academic qualifications.

The New Graduate Programs

The reform was much more successful with regard to creation of academic departments, research institutes, and graduate programs. Institutions like the universities of São Paulo, Rio de Janeiro, Minas Gerais, and Rio Grande do Sul—which had benefited in different degrees from past scientific traditions, the presence of foreign visitors, and opportunities of international exchange—could more easily adapt to and benefit from the new organizational formats. Small, high-quality graduate courses, which earlier hardly existed as an organized endeavor, were easily put together by these same persons and institutions.

This trend was reinforced by new sources of funding and institutional flexibility caused by the entrance of economic planning agencies into the field of science and technology. Suddenly and for several years, the amount of money available for science and technology far exceeded the ability to spend it. The new science and technology agencies created to handle these resources were flexible and modern; free from the bureaucratic and budgetary limitations typical of the Brazilian civil service; and looked with contempt at the complex, conflicted, and bureaucratized university institutions. They tried to stimulate research on Brazilian private and public firms, providing them with low-interest loans and technical assistance services. They soon realized that most of the competent people were in academic institutions and turned a large part of their resources over to them.

The strategy adopted by the science and technology agencies was to identify what they considered good or promising research groups and to provide them with direct support, very often bypassing the established procedures for labor contracts, accounting procedures, and decision-making within the universities. Substantive considerations were all that mattered. For the researchers, there was now a market that was sensitive to their qualifications and aspirations. For the universities, new resources became available, but they also flowed completely out of their control. Well-equipped, well-staffed, and well-paid departments and research programs began to exist side by side with poor programs—the first more concerned with research and graduate education, the latter bound to the traditional undergraduate schools and courses. A two-tier system was therefore introduced not only among higher education institutions but also within each of them, leading to tensions and ambiguities that would intensify in the years to come.

The education authorities had their own plans for faculty upgrading and graduate education. The new legislation required that faculty be hired and promoted only if they had the appropriate graduate degrees, and the universities were stimulated to create and expand their graduate programs. Quality was to be regulated through the Conselho Federal de Educação and an agency within the Ministry of Education, CAPES (Coordination for Improvement of Higher-Level Manpower), which predated the reform and was in charge of providing fellowships for faculty and graduate students within and outside the country.

For the planning agencies, their strategy worked very well. In 1970 some 57 doctoral programs were available in Brazilian universities; in 1985 there were more than 300, with another 800 providing training at the master's degree level.[8] About 90 percent of these courses were in public universities, and both levels combined were graduating about 5,000 students each year. By all accounts, Brazil had begun to build a significant scientific community.

8. Paulinyi et al. 1986.

There was only partial overlap between the policies of the educational agencies and the Science and Technology agencies. The premium placed on academic degrees led to a rapid proliferation of graduate programs throughout the country. The universities had the freedom to create them, and the teachers pressed for paid leaves to follow the new courses. Since most of the students were in "soft" fields, graduate education expanded most in those fields (see Table 11). In the process, quality usually suffered. The Conselho Federal de Educação, which was supposed to provide accreditation, was extremely slow and usually not very qualified to do the job. The agencies supported the programs of their liking, mostly in the basic and technological fields, and could not have cared less about the accreditation mechanisms devised by the council. Eventually CAPES established a peer review mechanism that became the de facto accreditation accepted by both sides. According to these evaluations, only about one fourth of the new graduate programs were of significant quality or had chances for improvement. This system was useful for allocating grants and fellowships but could not force a university to close down a program that did not qualify.[9]

High-Technology Institutions

The culmination of this drive was the establishment of brand-new institutions that would be free from the limitations of the past. They were to be as free as possible from institutional and bureaucratic limitations or restrictions; they were to receive large amounts of money from science planning agencies and put them to work in the hands of well-qualified people; and they were to work on the frontier of the modern technologies the country was supposed to need for its economic and industrial growth. Two institutions, more than any others, met these requisites: the Universidade de Campinas and the engineering program of the Universidade Federal do Rio de Janeiro, COPPE.

In 1965 Zeferino Vaz left the Universidade de Brasília to organize a new state university in the city of Campinas, São Paulo. At first the new university was to be mostly an institutional alternative to the Universidade de São Paulo, which was already too crowded with students and personnel. With Zeferino Vaz and heavy federal support, however, it became a project of a new and modern research university.

9. Castro and Soares 1986.

TABLE 11. Students Enrolled in Graduate Programs, by Field (1975–1983) (Thousands)

Year	Hard Sciences	Biological	Engineering	Health	Agriculture	Social, applied[a]	Social, humanities	Total
1975	2,898	2,196	2,421	2,111	1,811	10,808	—	22,245
1976	3,751	2,172	3,491	3,028	1,942	11,871	—	26,255
1977	4,362	2,405	3,969	3,370	2,374	15,052	—	31,532
1978	4,829	2,761	5,442	3,612	2,857	14,130	—	33,631
1979	4,755	2,951	5,459	3,771	3,018	16,654	—	36,608
1980	4,936	3,054	5,644	4,216	3,145	17,611	—	38,606
1981	5,170	3,137	5,715	4,677	2,709	18,776	—	40,184
1982	4,385	2,852	5,391	4,658	2,728	6,479	12,737	39,230
1983	4,264	2,913	4,990	4,561	2,709	6,452	9,961	35,850

SOURCE: Paulinyi et al. 1986.

[a] Social sciences and humanities were counted together until 1981; from then on, applied social sciences (social work, administration, communications, and so forth) are counted separately.

His conception for the new university was romantic. A new campus was to be built, and Vaz asked the architect to make "a large central plaza 300 meters across," and said:

> I will make it a wonderful garden, with the natural beauty of flowers, trees, stones, and water. This will be the Greek agora, and all units will converge on it.... The agora is attractive, and students and professors meet there to discuss and exchange ideas and concepts. You can find there the geneticist, the physicist, the physician, the botanist, the chemist, and the Faculdade de Engenharia de Alimentos. Multidisciplinary programs emerge everywhere, stimulated by the layout—because the circle provides a concept of unity: there are no privileged positions or sides. All converge on this plaza, which symbolizes the well-being of mankind.... I wanted to create a university like an organism in which the different organs—physicists, mathematicians, naturalists, philosophers, artists—all worked together for the preservation of the community's physical, mental, and spiritual well-being.[10]

Less romantic was his practice. Based on his reputation, and with financial support from the state and federal government, Vaz initiated an effort to bring back Brazilian scientists who had left the country in the previous years. In his interview Vaz told about his personal contacts and the support received from Finance Secretary Dílson Funaro at the state level and from powerful names in the federal economic and planning agencies, such as Minister of Finance Delfim Neto, Marcos Viana at the Banco Nacional de Desenvolvimento Econômico, José Pelúcio Ferreira at the Financiadora de Estudos de Projetos, and João Bautista Vidal at the Secretaria de Tecnologia Industrial.

Physics was to be a central activity, and Marcelo Damy de Souza Santos was invited to organize the institute, which was christened with the name of Gleb Wataghin. Cesare Lattes, who had left the Centro Brasileiro de Pesquisas Físicas, was also offered a position. Next came Sérgio Porto, followed by Rogério Cerqueira Leite, José Ripper, and several others. This group had in common old links with the Instituto Tecnológico da Aeronáutica and their years of work at the Bell Laboratories in the United States.

Sérgio Porto was to play a central role in the new project. He was born in 1926, studied chemistry at the Faculdade de Filosofia in Rio de Janeiro, and received a Ph.D. in physics from Johns Hopkins University in

---

10. Vaz interview.

1954. From 1954 to 1960 he worked in the physics department of the Instituto Tecnológico da Aeronáutica, and then he joined the technical staff of Bell Telephone Laboratories until 1965. From there he moved to the University of Southern California. A series of studies on the Raman effect based on the utilization of lasers led to many publications and an international reputation.

In spite of his achievements in the United States, in the mid-1960s Porto began to consider the conditions under which he would return to Brazil. In 1970 he came to Brasília to testify before a parliamentary investigative committee about the brain drain of Brazilian scientists. "I told them: The Brazilian people do not want us. We did not run away, you threw us out." For him, inadequate working conditions, not politics, were the reason so many Brazilian scientists lived abroad (his list included Sérgio Mascarenhas, Luís Valente Boff, Rogério Cerqueira Leite, José Ripper, Roberto Salmeron, and Fernando de Souza Barros). Several contacts with Planning Minister Reis Velloso followed, and in 1972 Porto agreed to return based on a pledge of $2 million for his projects.[11]

Sérgio Porto was supposed to become head of the physics institute, but when he arrived the position was already occupied by Rogério Cerqueira Leite.[12] Disputes over leadership and orientation led to Damy's dismissal, and Leite and Porto divided command of the new institution with Vaz. Eventually, the physics institute became what Leite described as "the largest physics institute in a university in the world"—and a good one in terms of quality—working in all areas related to semiconductors, from crystals growth to practical applications. Porto continued with his research on lasers and worked on ways to utilize them in a great variety of tasks, from eye surgery to atomic fusion. A company for technological joint ventures with the industrial sector, CODETEC, was also established under the leadership of Aldo Vieira da Rosa, an air force officer and scientist originally also from ITA.

While Zeferino Vaz was in charge, the Universidade de Campinas was

---

11. "This was his commitment right there near the swimming pool [at Porto's Los Angeles house]. He [Velloso] would tell Zeferino Vaz to provide the building and gave him assurance that he would hire thirty Ph.D.'s. These were my conditions in return: thirty Ph.D.'s, a building, and $2 million. And I got them. Unfortunately, I was naive, and the $2 million became just one. FAPESP had promised $300,000 but gave only 400,000 cruzeiros (about U.S.$67,000), which means that unfortunately I could not finish my laboratory" (Porto interview).

12. Cerqueira Leite had been a student of Porto's at ITA and went to work at the Bell Laboratories in 1962 after earning a degree in physics from the University of Paris. His return to Brazil was part of the same initiative reported by Porto and included also the promise of financial support and equipment.

considered to be in a provisional stage, which in practice meant that Vaz and his closest associates had full power to hire faculty and designate the university's authorities. This situation, combined with the extremely high proportion of "soft money" that became part of its budget, made the Universidade de Campinas an unstable, ambitious, and innovative institution, the closest Brazil ever came to a true research university.

While Campinas was planned as a classic university to which high technology was attached, COPPE was precisely the opposite—that is, an attempt to graft a high-technology teaching and research program to a traditional Brazilian university, the Universidade Federal do Rio de Janeiro.[13]

COPPE's history is inseparable from that of its founder and first director, Alberto Luís Coimbra.[14] Thanks to Coimbra's efforts, in the early 1960s Frank Tiller, his former professor at Vanderbilt (by then working at the University of Houston), was invited to teach at the Escola Nacional de Química. A series of visits by Brazilian chemical engineers to the United States followed, the goal being to establish a graduate program in chemical engineering in Rio de Janeiro. This interchange was supported in part by the Escola de Química, in part by American institutions, and in part by the Organization of American States. In 1962 Coimbra went to the United States to become more acquainted with the American model of graduate education. In the next few years he sent some of his best students—Giulio Massarani, Afonso da Silva Teles, Carlos Augusto Perlingero—to study in Houston.[15]

The project for a new institution was already being designed. Early in

13. For a full account, see Nunes, Souza & Schwartzman 1982.
14. Alberto Luís Coimbra graduated from the Escola Nacional de Química in Rio de Janeiro in 1946. He studied for a master's degree in chemical engineering at Vanderbilt University and from 1949 to 1953 worked at the Escola de Engenharia Industrial in São Paulo, a private institution. In 1953 he returned to Rio de Janeiro to compete for a chair at the Escola de Química. For the next several years he "taught at the Escola de Química, at Petrobrás [the Brazilian state-owned oil company], worked as a consultant for two American firms, taught at the Universidade Católica, and dealt with chemical engineering and the mechanics of fluids" (Coimbra interview).
15. "It was like a direct order from Coimbra. He was a very good teacher at the Escola de Química. People liked his courses, and it was a privilege to be treated well by him. I remember that Coimbra came into the library and gestured to me: 'Come here. You are going to Houston.' We believed in him so much that we did not think twice. . . . We had no idea what we were doing or what graduate education really meant" (Massarani interview). Massarani and Teles went to the United States "to get their M.A.'s and return to God knows what. If the project for a new graduate program went through, they would have a job, a career in graduate education at the university. If not, they would work in industry, where, at the time, they could do very little with the knowledge they obtained with their M.A.'s" (Perlingero interview).

1964 the project began to receive support from the recently created Fundo Nacional de Tecnologia and from the Banco Nacional de Desenvolvimento Econômico. This influenced its direction.

> A friendly relationship between professors and the officers at the BNDE was established. It was very gratifying, very pleasant. They shared the problems of the university and helped out when payment was delayed for some reason. The first bylaws for the fund were written at COPPE. At first they had only engineering, but we included physics, mathematics, and chemistry because one cannot do graduate education in engineering without the basic sciences.... We also helped establish the graduate program in mathematics at the Instituto de Matemática Pura e Aplicada.[16]

Money began to flow in 1967 and peaked in 1973. COPPE established a bewildering variety of cooperation agreements with persons and institutions all over the world. Its catalog for 1971 mentions the Organization of American States, the Fulbright Commission, the Rockefeller Foundation, the U.S. Agency for International Development, and the governments of France, England, the Netherlands, the Soviet Union, and Germany.

Through these agreements, foreign professors were invited, students went to get their doctoral degrees, and joint research projects began.[17] From chemistry, COPPE expanded into biomedical engineering, mechanical engineering, metallurgy, civil engineering, production engineering, nuclear engineering, naval engineering, urban planning, production engineering, and business administration. Several hundred professors were hired with better salaries and working conditions than those of the Escola de Engenharia, to which COPPE was formally attached. As the program's general coordinator, Coimbra moved to concentrate most decisions in his own hands. "We had to create almost an island to protect ourselves against the mold that surrounded us. We had to grow and to go up very quickly—like a balloon rising so quickly that no stone could reach us—to be strong and not tainted by the university's inefficiencies.... We

---

16. Coimbra interview.
17. "Coimbra was really concerned with getting all tendencies. Americans were somewhat prepotent in the organizations they supported—the OAS and others—and he did not like that because he felt restricted in his freedom.... Today COPPE has more European influences than American. But in some areas the Americans are stronger—for instance, in systems engineering. Europe and the United States are similar in terms of knowledge, and the Europeans are stronger in chemistry. But it is difficult to say; it is not clear" (Massarani interview).

had to make use of unorthodox means to do all that. This of course was not in accordance with the university's bureaucracy."[18]

Bureaucracy fought back. Coimbra was charged with mishandling public funds and left his post with bitterness in 1973. In 1977 the Universidade do Rio de Janeiro approved a new statute for COPPE that placed it under the direct authority of the university's rector. After the first years of strong entrepreneurial leadership, it was the time for management, and COPPE became a permanent and important component of the university.

How well did COPPE meet its initial goals? According to Coimbra, "COPPE was created to form a kind of professional Brazil did not have at the time, one at the master's and doctoral levels. We believed that people with these qualifications were necessary to the country's technological development. We had middle-level personnel, engineers, but we lacked graduate people who could create new technology."[19] This was precisely what the Fundo de Tecnologia defined as its own goals. According to the bank's economists, Brazil showed a clear lag between investments in basic sectors of the country's economy and investments in education. Industrial development, however, depended on the qualifications of the personnel, the strengthening of the country's scientific and technological competence, and the reduction of payments for imported technology, know-how, technical assistance, and patents. The fund was supposed to both invest in graduate education and provide incentives and technical assistance to stimulate Brazilian companies in their use of modern technology to meet the competition of the local branches of multinational corporations.[20]

The balance was not very positive. Coimbra, for one, showed his frustration.

> We created the graduate programs for a Brazil that did not exist and still does not exist, which did not correspond to what we expected to happen. We were throwing into the market a sophisticated product meant for the country's technological development. We imagined that if we did our part in forming creative people in engineering, they would be absorbed by a country that really wanted to create its own technology. But it never happened.... Brazil does not need M.A.'s and doctors, not even five-year engineers. Operational engineers are enough, since we will keep handling imported factories forever.[21]

18. Coimbra interview.
19. Coimbra interview.
20. BNDE 1974.
21. Coimbra interview.

Other opinions were more balanced but pointed in the same direction:

> I am almost sure that COPPE is full of defects, in the sense that it goes too far ahead of Brazil's reality and is too sophisticated with regard to the practical side of the productive activities in many fields.... The industries were completely resistant to any participation of this kind, now or in the past. Our industrial development was based exclusively on the importation of foreign technology, on multinational corporations, or on the acquisition of foreign patents by Brazilian companies.[22]

A more precise picture can be gathered from an analysis of data on student graduation and their future work (see Table 12). Between 1964 and 1978, only twenty-five students received doctoral degrees. Only about 20 percent of the master's students ever got their degrees, and 50 percent abandoned the courses without getting their credits. These low rates of completion and graduation are similar to what is found in most graduate courses throughout Brazil. In that sense, COPPE is not exceptional, but it is obviously no less troublesome.

If we leave aside the more recent courses, it is obvious that the highest rates of desertion were in fields where enrollment was also highest. These figures suggest that, for most students, graduate education was just a way of extending somewhat their student life, most of the time with a fellowship, while they waited for a place in the job market. These figures can mean also that COPPE students were in such high demand that they had no time to finish their degrees.

Either way, it is obvious that the level of education offered by COPPE was much higher than what was demanded by the students who do not get their degrees or by the companies that employed them. The same or better results could be obtained if COPPE were geared to provide more modest levels of training for the bulk of its students and concentrated its effort in graduate education on the 20 percent or so who finish their degrees. For those who do get their master's degrees, the destination is the university itself (37 percent) or public employment (21 percent). Only 13 percent enter the private sector, while 19 percent continue their student life at the doctoral level. Those who became teachers after getting a master's degree probably worked with undergraduate students and were not able to continue a professional life geared toward research and technological development. Desertion rates for those who work for their doctoral degrees was still higher than at the master's level.

22. Pinguelli Rosa interview.

TABLE 12. COPPE: Graduation, Enrollment, Desertion, and Destination of Students, 1965–1978

Field	Ph.D.'s	Master's Degrees	Master's Enrollment	Desertion	Destination[a]
Chemical engineering (1963)	4	124	437	52.6%	Teaching, public companies
Mechanical engineering (1966)	3	66	268	48.5%	Teaching, private companies
Electrical engineering (1966)	1	86	699	60.9%	Public companies, teaching
Metallurgy (1966)	2	59	368	30.2%	Teaching, public companies
Civil engineering (1967)	6	139	942	65.5%	Teaching, doctoral programs
Production (1967)	3	140	852	68.9%	Public companies, teaching
Naval (1967)	—	25	97	43.3%	Teaching
Nuclear (1968)	—	87	329	44.3%	Doctoral programs, teaching
Systems engineering (1971)	6	109	660	49.1%	Teaching, doctoral programs
Biomedical (1971)	—	21	120	33.3%	Teaching
Business administration (1975)	—	12	160	20.6%	Public companies

SOURCE: Nunes, Souza & Schwartzman, 1982:241–42.
[a]Main occupation of those who obtained their master's degrees. Two occupations are given when the figures are close.

In short, COPPE seemed to become a good program of specialization in the most traditional and operational fields of engineering: civil, electric, systems, and production engineering. Its original goals—to develop a truly national and internationally competitive engineering capability—were mostly frustrated. The ideals of research, academic work leading to dissertations, a constant flow of professors to provide the highest possible levels of technical and scientific competence, the permanent interchange with Europe and the United States—these notions, what many call "the COPPE spirit," contrast sharply with the reality: the desertion of 80 percent of its students and the employment patterns of the others.

Big Science and High Technology

The whole rationale for investment in technology is its practical utilization, and a survey of experiences of this kind would have to include the whole field of agricultural research; recent achievements in biotechnology, the airplane and weapons industry, and the technology for steel production;[23] and the role of such institutions as the Instituto de Pesquisas Tecnológicas in São Paulo. Such a survey would be beyond the scope of this study, but two extreme cases of high technology—atomic energy and computers—stand out from the others, in part because of their proximity with the basic sciences, and require a closer look.

We have seen how research in advanced physics, which provided the basis for some applied work during the war, could not perform the same feat afterward in the much more complex, expensive, and politically charged field of nuclear technology. Confronted by the alternatives of trying to develop its own technology with the help of the existing scientific community or acquiring foreign technology, the Brazilian government chose the second. In 1975 an ambitious agreement for nuclear cooperation was signed with West Germany, which implied the construction of several nuclear energy plants and the transfer of enriched uranium technology. The agreement drew strong opposition from Brazilian scientists, because it consisted mainly of the transfer of engineering technology and did not incorporate the acquired or presumed competence of Brazilian scientists. In time the agreement proved to be overambitious, and it is now limited at most to the construction of two power plants, neither of which is near completion at this writing (the Westing-

---

23. Dahlman and Fonseca 1987.

house plant, meanwhile, is plagued by successive technical difficulties and may have to be scrapped before going into full operation).

A so-called parallel program of atomic research was also undertaken by the Brazilian military, outside the restrictions built into the German-Brazilian agreement. Rumors that Brazil is developing its own atomic bomb have never been confirmed; still, the government has acknowledged development of nuclear engines for ships and submarines, and in September 1987 it was formally announced that Brazil had developed all the needed technology for the production of nuclear fuels for peaceful purposes. The method of ultracentrifugation was said to be similar to the one used by the URENCO consortium in Europe, and the grade of enrichment, which was announced to be of 1.2 percent, was supposed to increase to 20 percent in one or two years, when an industrial plant was supposed to begin working.[24] The work had been going on for eight years at the Instituto de Pesquisas Nucleares at Universidade de São Paulo with support provided by the navy and at a stated cost of $37 million. The announcement was received with widespread skepticism by the Brazilian press and Brazilian scientists. The level of enrichment was considered too low for any practical purposes and the expenses too high, given the country's deep economic crisis and the other needs of scientific institutions. The concentration of resources in military research at the expense of civilian institutions was seen as a worrisome trend.[25]

While a key feature of the nuclear programs was the exclusion of university-based scientists and the creation of large, state-controlled bureaucracies, the computer industry began with people coming out of the universities and gave rise to a large number of privately owned companies. The origins of this industry can be traced back to the physicists and engineers trained by the Instituto Tecnológico da Aeronáutica. Later it drew from those educated or engaged in research in such places as the Universidade de Campinas, the Escola Politécnica of the Universidade de São Paulo, the Universidade Católica in Rio de Janeiro, and COPPE. At the end of the 1960s the Banco Nacional de Desenvolvimento Econômico began to support research and development in computer science and microelectronics. In 1971 the navy contracted with the Universidade

---

24. The entire nuclear program was put in disarray with the economic crisis of the late 1980s, and the industrial plant never became operational. In 1990 the Fernando Collor government sealed a deep well drilled by the military for underground nuclear testing and signed an agreement of mutual inspection with Argentina to ensure that the military side of the nuclear testing would stop.

25. Guilherme 1957; Sales 1958; H. G. Carvalho 1973; Gall 1976; Leite 1977; Morel 1979; Adler 1987; *Jornal do Brasil* 1987.

de São Paulo and the Universidade Católica in Rio de Janeiro "for the planning, development, and manufacture of a computer prototype suitable for naval operations, preferably in association with Ferranti," an English company.[26] In 1972 the government created an agency to coordinate the whole area of computer acquisition and data-processing for the public sector, but with powers to control imports for the whole industry. This commission led to the creation of the Secretaria Especial de Informática, established in 1979 under the National Security Council with full powers to decide everything related to computers and microelectronics, from imports to the establishment of manufacturing firms or the acquisition of data-processing equipment by universities or state corporations. In 1973 a state-controlled holding, Digibrás, was established with capital from state-owned corporations to promote a Brazilian computer industry, and in 1975 it gave rise to Cobra S.A. With these instruments, a policy of market reserve for the production of microcomputers for Brazilian firms was established. For minicomputers, the strategy was to stimulate joint ventures under the control of Brazilian firms and a full transfer of technology; the market for mainframes remained open to IBM and other multinational firms, but all their actions came under scrutiny.

This policy coincided with the worldwide explosion of the microcomputer industry, and in Brazil it started with imported components and was patterned on well-known products manufactured by Sinclair, Tandy Corporation, Apple, IBM, and their clones. Research was mostly on reverse engineering, software adaptation, and the development of less-complex circuitry and parts. In 1983 Brazilian computer firms, which had barely existed five years before, were employing about 16,000 people and selling about $690 million in equipment, while multinational firms in the country were employing about 10,000 people and had sales of $800 million.[27]

By the end of 1984, a bill defining a "national policy for informatics" was approved by the Brazilian Congress by large majorities from the government and opposition parties. It was preceded by a lively debate reminiscent of the one leading to Brazil's state oil monopoly in the early 1950s. Once more the issue was presented as a dilemma between national autonomy and self-determination, on the one hand, and control of the country's economic resources by international companies and their local associates, on the other hand. Again the proposed policy gathered

---

26. Adler 1987:245.
27. Secretaria Especial de Informática 1984.

intense support from those ranging from the nationalist military groups to intellectuals, students, scientists, trade unions, and politicians.[28]

The many remarkable elements in this policy generated several studies, international press coverage, and threats of retaliation from the Reagan administration. The main novelty was that this was the first attempt in Brazil's history to develop an industrial policy based on local technology and purely Brazilian firms. Market protection had existed for many years, for instance, for a small group of multinational corporations in the automobile industry; and state monopoly for oil had also existed for many years, but based on internationally available technologies. Here, however, some level of technological competence had been previously built, and it was to provide the foundations for the new policy.

The Brazilian policy for computers is an extreme case of "backward integration," in which production starts with assembly of the final product with imported parts, with the expectation that the product will incorporate an increasing amount of locally produced components.[29] Such a policy requires, among other things, a corresponding investment in research and development, which in the Brazilian case does not seem to have existed.

A picture of the current stage of university research in computer sciences (which does not include such related fields as electronics or automation) can be gathered by examining a three-year research plan put forward by the Sociedade Brasileira de Computação, a scientific association, and the Centro Tecnológico de Informática, a research outfit created by the Secretaria Especial de Informática near the city of Campinas. There are five institutions providing doctoral degrees for about ten people a year, and the total number of researchers with doctoral degrees is 108. There are also fifteen institutions providing master's degrees and a small number of people being trained abroad and returning to the country. At most, fifteen new Ph.D.-level researchers are expected to enter the field each year. The total number of researchers outside industries is estimated at 750, five hundred of whom work in software. The research plan projected an increase in the number of Ph.D's to 750 by the year 1997, against 300 if the current growth rate stands. For this, proportional increases in technical personnel, equipment, library facilities, and so forth will be needed. The total cost for all research projects, infrastructure, interchange, and follow-up was estimated at about $40 million in three years, a very small amount if com-

28. What follows is based on Schwartzman 1988b; see also Tigre 1983; Piragibe 1985; Frischtak 1986; Evans 1986; Adler 1987.
29. Nau 1986.

pared with what is being invested in the developed countries, and close to the value of the equipment IBM was willing to provide Brazilian universities in the same period.

A list of high-priority research projects was also drawn up. It included the design of digital systems, time-sharing systems, software engineering, data bases, CAD/CAM, artificial intelligence, sign processing, mathematics applied to computing, and computer theory. In all, sixty-five projects were identified. The research plan was an aggregation of existing research projects, to which a weak order of priorities was attached. The plan was never funded as it stood.

There is no equivalent information for research in industries, but its scale could be inferred from the existence of about 4,000 employees with university diplomas in all Brazilian companies, most of them working on sales, maintenance, quality control, and administration. The Centro Tecnológico de Informática was supposed to become the starting point for a large research and development establishment, to be supported with a special tax and placed directly under the authority of the Secretaria de Informática. The tax was vetoed by President João Figueiredo from the 1985 law, and the Centro remained as a small outfit, with about 300 people and a budget of around $1 million a year. It is not a purely research institution, since it is supposed to sell services to the private sector, to provide technical assistance, and to develop joint projects with the universities. It is also responsible for overseeing the fulfillment of nationalization targets of IBM's computer assembly plant nearby (geared only to the foreign market) and for developing standards and providing certification for the national computer industry.

The weakness of the research effort is one reason the Brazilian policy for the computer industry came under severe criticism, not only from foreign competitors who would like to sell in the Brazilian markets but from end-users and manufacturers in Brazil who see this policy as a growing roadblock to their access to state-of-the-art technologies. In computers, as in atomic energy and in other applied fields, the great leap forward was far shorter than originally expected.

10

EPILOGUE

Freshness and Decay

Cities in the New World, wrote Claude Lévi-Strauss, go from freshness to decrepitude without ever maturing.[1] He might have withdrawn the statement during his 1984 visit to the city on the occasion of the Universidade de São Paulo's fiftieth anniversary. As one of the largest and busiest industrial cities in the world, São Paulo does not seem to have grown in the directions he may have sensed on his first visit, for the inauguration of the Faculdade de Filosofia. And yet he was probably right in a deeper and unexpected sense.

The creation of new universities, the quest for excellence, the organization of research programs, the drive for technological self-reliance, the concern with practical utilization of scientific knowledge to meet social and economic needs—all suggest an element of freshness, youth, and dynamism that have always been present in the growing scientific com-

---

1. Lévi-Strauss 1955, opening chapter on São Paulo.

munity in Brazil since Bonifácio de Andrada. Maturity, however, does not follow naturally from youth, just as reality does not follow easily from wishes or ideology.

There is much more science and technology in Brazil today than only twenty years ago, but it is clear that a space for science, in terms of socially defined, accepted, and institutionalized scientific roles, is barely there. At most there are islands of competence, niches where science could develop for some time, but always precariously and threatened by an unfriendly environment.[2] In social life, decrepitude rather than maturity usually follows failed institutionalization. Decay occurs when founders of scientific institutions get older and are unable or unwilling to open space to new ideas, new generations, new leadership; when generous and ambitious projects of social reform, like the scientific movements of the past, are gradually transformed into barely disguised ideologies for the protection of narrow interests; when education loses its role of expanding opportunities and competence and becomes a mechanism for social inequality and privilege. The passage from freshness to decay is ambiguous and often difficult to grasp, since nothing changes—the persons, the institutions, their discourse—except their contact with reality and their premature aging.

The crucial question about the Brazilian scientific community today is whether it is heading toward maturity or whether the bleak prediction of Lévi-Strauss is coming true. As of this writing the ghost of premature decay is very present in Brazilian society in a context of acute economic crisis, a shaky institutional order, and a society marked by the entrenchment of narrow interest groups and lack of consensus on basic values. This situation affects the scientific community as it affects all organized sectors in Brazil. It is very difficult, but necessary, to try to understand whether the present malaise affecting the scientific community is due only to circumstantial and external factors, such as the debt crisis and the political instability generated by the political transition to democracy, or whether it is more structural and therefore less likely to go away.

The uncertainties of the present dramatize the fact that, in spite of the achievements of Brazilian science in the last fifty years, its place in society is still far from recognized. We have seen how, in the past, Brazilian science flourished only under the protection of the powerful, whether Emperor Pedro II or Planning Minister Reis Velloso; under the guise of applied technology, as in the Instituto Manguinhos; or in atypical institutions, such as the Faculdade de Filosofia e Letras of the Universidade de São Paulo. From 1945 to 1964, the last time Brazil enjoyed an open

2. J. B. A. Oliveira 1984.

political system, the scientific community was too small to have any socially significant presence, and its main institutional achievement of the period, the creation of the Conselho Nacional de Pesquisas, was in fact the outcome of an extremely elitist and frustrated project of nuclear development.

Brazilian scientists have always sensed the weak links they had with the broader society, and they often looked for escape routes in politics, education, and the economy. Almost all, and many of the brightest, looked for political participation outside their laboratories, taking part in movements for university reform or looking for applied subjects for their work. In an open political regime these tendencies are likely to reappear and intensify. We can conclude this journey through the formation of the scientific community in Brazil by looking briefly at these broader contexts as they stand today.

Politics: From Military to Civilian Rule

Political scientists still discuss why the Brazilian military stepped down in 1985 in favor of civilian rule. They still ponder the broad consequences of this peculiar type of transition.

Transition to civilian rule was, from the beginning, part of the stated purposes of the Brazilian military elite, the group of high-ranking officers that came to be known as the "Sorbonne." They had a task to perform, modernization of the country, and then a period of "slow, gradual, and secure" transition to democracy was to take place. What happened can be better described as an example of strategic retreat. Some of their ambitious projects did come to fruition—typically those physically isolated from the large urban centers, endowed with heavy investments, based on established technologies, and often related to powerful interest groups—such as the Itaipu hydroelectric plant in the South, the Carajás mining complex in the North, the development of an industry of military equipment for export, the alcohol program for engine combustion, and the military project of uranium enrichment. Some of these projects are questionable in terms of their cost, environmental impact, and sheer morality, but at least they do exist. They are, in any case, exceptions. From 1978 on, as the availability of foreign loans dwindled, the price of foreign oil soared, and the political basis of the military regime narrowed, most of the big projects of the previous years collapsed: the nuclear energy program, the steel railroad in Minas Gerais, the big trans-Amazon and Rio-Santos roads, the beginnings of a Brazilian naval industry, and, more markedly, all promises of urban renewal, social assistance to the poor, rural modernization, and reform.

We have seen how Brazilian science also grew in the wake of the 1970s. In terms of scientific production reflected in the *Science Citation Index,* Brazil was thirty-first in the world and fourth in the Third World in 1973, with about 0.25 percent of the articles published worldwide. By 1978 Brazil was already second in the Third World, after India, and twenty-fifth worldwide. The number of articles published in the international literature went from 812 in 1973 to 1,060 in 1978 and 1,551 in 1980. This was a significant increase in national and regional terms, but it did not change much in terms of international weight. The worldwide pattern of concentration of scientific production is reproduced within Brazil. Five institutions—the Universidade de São Paulo, the Universidade do Rio de Janeiro, the Universidade de Campinas, the Universidade Estadual Júlio de Mesquita (São Paulo) and the Escola Paulista de Medicina—produced in 1982 some 43 percent of all articles, books, and scientific communications surveyed by CAPES; 43.6 percent of all articles published in international journals; and 70 percent of all articles listed in the *Science Citation Index* for 1973–78.[3] After the expansion period, Brazilian science was like the rest of the country: there were many significant achievements, many halted initiatives and projects, and a pervasive uncertainty about the future. The failure of the great leap forward is usually explained away by an array of unfortunate circumstances. The standard reasoning is still that Brazil was just unlucky, victimized by the increase in oil prices, soaring interest rates, and the fall in international commodity prices. The military is also blamed. They are accused of being clumsy and authoritarian and of not taking advice and direction from the scientists. The lack of an open political regime precluded close public scrutiny of how well and for what purposes resources were being spent. As the opposition to the military regime mounted, so did hopes for civilian rule: the new regime would be open and attend to the needs of the people, would not submit to the whims of the International Monetary Fund, would shut down the purely military and technocratic projects, would place the science and technology agencies under the control of the scientific community, and would make space for a real reform of Brazilian universities. Brazilian scientists have always longed for a Ministry of Science and Technology, and so the ministry was created. Those in theater, movies, and the performing arts complained about the lack of support for cultural activities and were also granted their minister. Because people complained that the military had spoiled

3. C. de M. Castro 1986a; Morel and Morel 1977; Garfield 1983.

the universities, a national commission to examine what was wrong and to decide what to do was to be created. The new regime would be open to all, and nobody would be left out.

In hindsight, it is obvious that the establishment of a Ministry of Science and Technology and the occasional statements by public authorities about their commitment to science, technology, and higher education were not enough to provide the Brazilian scientific community with the space, recognition, and support it expected to receive from the new regime. The newly created Ministry of Science and Technology did not make any contribution to changing this picture in any significant way. In terms of orientation, as in terms of many of its key figures, the new ministry tried to be a carrier of the old ideals of science planning, technological nationalism, and self-sufficiency. But it was born very weak and brought together under its umbrella only previously existing institutions, such as the Conselho Nacional de Desenvolvimento Científico e Tecnológico and FINEP, to which was added the Secretaria Especial de Informática. Most of its resources went to space research, computer science, and biotechnology.[4] Nuclear research, agriculture, industrial technology, military research, and university research all remained outside its scope. The ministry was perceived as a government's political concession to some sectors of the PMDB party and the scientific community, and the split that existed in the past between technological and economic policies or decisions was reproduced now amid fewer resources and more isolation than during Reis Velloso's tenure at the Ministry of Planning.

The main traits of President José Sarney's government are its lack of any long-term project or commitment (except to its own survival as a political arrangement) and its extreme susceptibility to pressures of well-organized and vocal interest groups. More to the point, it is now clear that the political enlightenment inherent in an open political regime does not necessarily imply an equally progressive approach to matters of science, technology, and education. Openness to interest groups and public opinion is of course a desirable feature in any democracy, but it creates special problems for a scientific community not used to fighting for its own space. It is only natural, under the circumstances, for scientific groups to approach the political parties and attempt to bring to the parties' programs and projects a proper evaluation and acknowledgment of their work. The risk is the overpoliticization of scientific life, with the prevalence of political and ideological criteria over those of scientific and intellectual competence in the scientific community's leadership.

---

4. Ministério de Ciência e Tecnologia 1985.

Mass Higher Education

We have seen that science in Brazil developed mostly around a few academic institutions. In spite of emphasis given to technology in the last twenty years, most of the research that exists today is carried on in Brazil's best universities and academic centers. Yet recent developments suggest that this space, conquered with difficulty in a few decades, is far from secure.

In the 1970s, higher education in Brazil drifted further and further away from the unified university research model prescribed by the 1968 reform. In 1985 it became an extended, complex, and highly differentiated system with the following main features:[5]

- A small elite of about 14,000 faculty with doctoral degrees or equivalent titles (sometimes known as the "high clergy" of Brazilian education) and about 40,000 students in master's and Ph.D. programs in the best public universities, mostly in the southern part of the country. Professors are endowed with reasonable salaries and can complement them with fellowships, research money, and better working conditions (in spite of declining resources in the 1980s). Students are selected from the best coming out of the public universities, do not pay tuition, and get fellowships for two or more years.
- About 45,000 full-time teachers with relatively low academic status (sometimes known as the "low clergy") serving about 450,000 students in free, public universities throughout the country. Hired initially on a provisional basis, without formal procedures or evaluation, most of these teachers are now tenured and can be promoted up to the assistant professor level by seniority. Courses and facilities at this level are uneven, with the best in the center-South and in the traditional professions, and the worst in public universities of the Northeast. Faculty is mostly full-time (or at least paid on a full-time basis), and members seldom have more than a bachelor's degree. Students have access to almost-free restaurants and a few other facilities, but lodging is rarely provided, and physical installations, laboratories, research materials, and teaching aids are scarce. Students usually come from the best private secondary schools—which means they are from middle- to high-income families—and often go through cram courses to prepare for the university's entrance examinations. These courses are provided by private, profit-oriented firms outside any kind of govern-

---

5. Schwartzman 1988a.

ment supervision and tend to be very efficient for their purposes. Thanks to loans provided by the Inter-American Development Bank, most federal universities built their brand-new campuses on the outskirts of the cities where they are located. There were few provisions for housing, since the government was afraid of too much student concentration, and in any case Brazil lacks the tradition and resources to move students to different locations to study on a large scale. Today most of these campuses are poorly maintained and inconvenient to use. Those that can, try to remain in their old seats in central urban areas. As education opportunities expand, students face increasingly serious problems of unemployment, in spite of the relative quality of their education.

- Around 60,000 teachers serve about 600,000 students in private institutions. Most of these teachers work part-time, are not well qualified, and must carry a large teaching load in several institutions—or a combination of jobs—in order to survive. Some have full-time appointments in public universities and moonlight in private schools, where courses are usually taught in the evening. They are not organized and do not have the same teachers' associations prevalent in the public sector. Tuition is low and government-controlled, but students can barely afford it. Facilities and teaching materials are minimal or nonexistent. Students tend to be poorer and older; courses are mostly in the "soft" fields. Most students are already employed in lower-middle-class or white-collar jobs and look for education as a means to job improvement or promotion; they are usually more interested in credentials than knowledge or skills.
- A profound regional imbalance exists, contrasting the southern states, and more specifically the state of São Paulo, with the rest of the country. São Paulo is Brazil's biggest and most industrialized state, encompassing about one-fifth of the national population and one-third of the national graduate enrollment. This is also the region where the dual nature of the Brazilian higher education developed more fully. There is proportionally less enrollment in universities than in other regions, but the state universities are better than in the rest of the country, while the private sector is much more complex and differentiated than elsewhere. There are just three federal institutions in São Paulo, a small university in the city of São Carlos, the Escola Paulista de Medicina, and the Instituto Tecnológico da Aeronáutica in São José dos Campos. This is in contrast to the country's poorest region, the Northeast, where more than 70 percent of the students are enrolled in federal universities, whose academic standards are usually lower than those in the South.

In 1985 the new government formed a national commission to make recommendations for a new reform of higher education institutions. The commission's suggestions included the establishment of higher levels of differentiation, autonomy, and accountability among public institutions; public support for private, high-quality schools and universities; the introduction of new modalities of public education for older and working students; and increased support for scientific research from within the Ministry of Education on a competitive basis.

These recommendations were never put into effect, mostly because of opposition from teachers' associations. In last analysis, demands for accountability, effective autonomy, and quality come mostly from the more academic groups in higher education institutions, which feel they are losing their space in an increasingly unionized and politicized university environment. These groups, however, can look for support elsewhere—in research funding agencies or in the private sector—and are not likely to carry their demands for academic improvement too far. The only sectors completely locked into the higher education institutions are the so-called low clergy and the university bureaucrats in public institutions. Their professional qualifications are usually not good enough to allow them to move easily to comparable jobs in the private sector; they have no way to raise additional resources through research projects; and they are often located in regions with very limited middle-class job opportunities. Well-organized and politicized, this sector has been able to get the government to provide job tenure, promotions based on seniority, and fixed and homogeneous salary scales.

This scenario points to the threat of a progressive "Latin-Americanization" of Brazilian federal universities, with the alienation of its more competent sectors, and the progressive politicization of its daily life. One possible trend to be expected is the continuous lowering of admission standards and spreading of night courses in the public institutions, followed by migration of the richer and better-educated population to private institutions. Research money can follow the same path or remain concentrated in nonuniversity government institutions.

The expansion of higher education into a highly stratified mass system brings additional tensions and difficulties to the scientific community. The high salaries paid, the facilities for travel abroad, the use of English as a language for publication, the preference for research over teaching, and the choice of research topics that are intellectually attractive and prestigious, rather than useful and practical, are often seen by the scientists themselves as unjustified privileges that reinforce the scientists' elitism and help maintain the present patterns of regional imbalance, underdevelopment, and economic dependence. These feelings of inadequacy and social inequity are compounded by the difficulties of carrying on successful scientific lives in an unfavorable environment. Not surprisingly, this combination often leads to futile attempts to abandon the enterprise of modern science altogether and to look for different paths. Would it not be possible to find a science that is closer to the poor, expressed in ways and in a language that all can understand? Shouldn't we distribute research resources in a more equitable and regionally balanced way, rather than following the always questionable standards of merit? Shouldn't scientists renounce the intellectual games of the rich and look only for knowledge that is obviously useful, cost-effective, and practical?

Technology and Economics

In developed countries most of the resources for scientific and technological research are spent on applications; in developing countries the opposite seems to be the rule.[6] Brazilian plans for science and technology, and the behavior of its science and technology agencies through time, show an attempt to approach the pattern of expenditures of developed countries. So far there is no clear evidence that these efforts to develop applied knowledge and bring it to industry have produced substantial benefits. Brazilian industrial firms, public and private, have usually shown little interest in original research and development. The Brazilian pattern of economic growth has always emphasized the free admission of foreign capital, enterprises, and technology. Given this situation, it is reasonable to ask whether all efforts to develop indigenous technology are not being wasted.

An implicit assumption of many investments in technology is that if good products or processes can be obtained they will somehow become socially or economically beneficial. Experience seems to show that this is not always the case. Technological research within a university or a research institute is relatively inexpensive and inconsequential. Adoption of a product in the marketplace is an altogether different matter, requiring much higher investments and serious consideration of the underlying and expected market conditions. However, the presence of competent technologists and a reservoir of technical expertise may open alternatives that would not otherwise exist; the strengthening of scientific and technological staffs at the universities might lead to creation of institutional bridges between academic research centers and industry, and they can

6. Moravcsik 1975:108.

always improve the country's pool of educated workers, an important asset in itself.

The reasons behind the efforts toward indigenous technology are far from simply economic; they include such considerations as national pride, employment for technologists, development of skills and technological confidence, creation of demand for suppliers of parts, and strengthening of sectors in the public bureaucracy.[7] Political motives increase the desire to keep all aspects of technology that are important for national security and independence—including communications, energy, computers, and military equipment—within the country. Economic considerations are usually long-range and roundabout; they are based on the hope that, in the long run, the costs of technological licensing will be higher than the costs of technological self-sufficiency.

These social and political considerations are based on the view that market mechanisms cannot bring an underdeveloped country to a state of satisfactory economic development and social justice. Foreign firms operating in underdeveloped economies usually come with their technologies fully developed and train their workers only in routine procedures of operation and maintenance. Indigenous firms prefer to buy well-tested machines and processes abroad, which usually include contracts covering replacements and technical assistance. Imported technology is also usually laborsaving and produces sophisticated goods for the wealthier classes, leaving large sectors of the population without provision.

The solutions to these problems are not obvious. Policies that aim at technological self-reliance easily lead to the protection of inefficient industries; to the maintenance of expensive, unproductive, and poor research; and to the growth of costly and sluggish bureaucracies.[8] The inefficiency of the state, in contrast to the economic rationality of private corporations, is often cited as the reason behind the difficulties of the projects of technological self-reliance in countries like Brazil or India, in contrast to the achievements of market-oriented economies like South Korea or Taiwan. These, however, are clear examples of garrison states that have much broader powers than the extended, amorphous, and contradictory state bureaucracies of the former and that developed comprehensive policies in which technology is fostered not in a free market but with full consideration of its market and long-term implications.[9]

7. Erber 1977.
8. Wade 1985; Bauer 1977.
9. Nau 1986:14.

These countries also had a pool of well-educated and disciplined workers that does not exist in other societies.

The Demon

This pessimistic overview, following a long survey of the efforts to build an effective space for science in Brazil, brings into question the effective value of this whole drive. And yet it is the very existence of this effort and its constant renewal through the years, rather than the eventual success or failure of specific projects and undertakings, that opens the space for hope.

I began with the myth of Sisyphus, and I may as well end with a similar image. To find its space, scientific research must assert its own worthiness, independent of its broader implications and consequences for Brazil's educational, technological, and economic institutions. Ultimately, the question is not whether science is accessible or inaccessible to the people, useful or not for technology, pertinent or not for national pride and grandeur. The question is whether there is a consensus, at least among a significant number of people, that Brazil should become a modern country and participate in the common fate of our time, which requires a systematic effort of self-clarification and knowledge in an increasingly rationalized and intellectualized world—and whether this consensus can be maintained and expanded. The history of the scientific community in Brazil shows that, in spite of all the difficulties, there is today in the country a sizable and growing group of people committed to these values, which is a reason for optimism. The commitment to the expansion of human knowledge and competence, as Max Weber said many years ago, is ultimately a matter of choice, a value judgment that cannot be either demonstrated or refuted by its practical utility or short-term consequences—even if we believe, as we do, that these consequences can be of great significance for all. If we agree on these values, "we shall set to work and meet the 'demands of the day,' in human relations as well as in our vocation. This, however, is plain and simple if each finds and obeys the demon who holds the fibers of his very life."[10]

10. Weber 1958:156.

APPENDIX:
LIST OF INTERVIEWS

The following interviews are from FINEP's project on the social history of science in Brazil. The texts of most are available at the Centro de Pesquisa e Documentação em História Contemporânea do Brasil, Fundação Getúlio Vargas, Rio de Janeiro. Interviews not connected with the FINEP project are listed at the end of this appendix under "Other Interviews."

Ab'Saber, Aziz Nacib. Geographer. Graduate of the Faculdade de Filosofia, Ciências e Letras, of the Universidade de São Paulo. Professor of physical geography at the same university.

Amaral, Afrânio (1894–1982). Specialist in public health and tropical medicine. Director of the Instituto Butantã from 1928 to 1935.

Beck, Guido (1903–1989). Physicist. Born in Austria, worked in Vienna, Leipzig, and in England at the Cavendish Laboratory at Cambridge University, among other countries. Worked at the National Astronomic Observatory in Cordoba, Argentina, between 1943 and 1951, and since then at the Centro Brasileiro de Pesquisas Físicas in Rio de Janeiro.

Bier, Otto G. (1906). Microbiologist. Worked at the Instituto Biológico, the Instituto Butantã, and the Escola Paulista de Medicina.

Braga, Ernani (1913–1984). Specialist in sanitary medicine for the Instituto Oswaldo Cruz. Medical doctor for the Universidade do Brasil (1935). General director of the Departamento Nacional de Saúde (National Health Department) in 1954. Director of the Division of Human Resources, World Health Organization (1967–1973). Active in many teaching and training activities in the field of public health.

Brieger, Friedrich Gustav (1900). Geneticist. Born in Germany, holds a doctorate from the University of Breslau. Came to Brazil in 1936 to teach genetics at the Escola Superior de Agricultura Luiz de Queiroz in Piracicaba.

Castro, Almir (1910). Specialist in public health. Director of the Campanha de Aperfeiçoamento de Pessoal de Nível Superior (CAPES) between 1954 and 1964. Vice-rector of the Universidade de Brasília with Anísio Teixeira.

Chagas Filho, Carlos (1910). Biophysicist. Researcher at the Instituto Oswaldo Cruz since 1935. Founder of the institute of biophysics of the Universidade Federal do Rio de Janeiro.

Coimbra, Alberto Luís (1923). Chemical engineer. Founder and first director of the Coordenação de Programas de Pós-Graduação em Engenharia (COPPE), Rio de Janeiro.

Cordeiro, Antônio (1923). Geneticist. Studied with Dobzhansky at Columbia University. Worked at the Universidade de Brasília, Universidade do Rio Grande do Sul, and later the Universidade Federal do Rio de Janeiro.

Costa Neto, Cláudio. Chemical engineer. Worked at the Universidade Federal do Rio de Janeiro. An assistant to Fritz Feigl at the Laboratório da Produção Mineral in Rio de Janeiro.

Damy de Souza Santos, Marcelo (1914). Physicist. Was director of the Instituto de Energia Atômica in São Paulo and president of the Comissão Nacional de Energia Nuclear.

Danon, Jacques (1924–1989). Physicist. Studied with Irene Joliot-Curie at the Radium Institute in Paris. Researcher at the Centro Brasileiro de Pesquisas Físicas in Rio de Janeiro.

Dias, Mário Ulysses Viana (1914). Physiologist. Worked at the Instituto Oswaldo Cruz with Miguel Osório de Almeida. Professor at the Universidade Federal Fluminense in Rio de Janeiro.

Djerassi, Carl (1923). Specialist in chemistry of natural products from Stanford University. In the 1950s developed a series of joint projects at the Instituto de Química Agrícola in Rio de Janeiro. Between 1969 and 1976 presided over the program of scientific cooperation between the American Academy of Sciences and the Conselho Nacional de Pesquisas.

Duarte, Paulo (1899–1984). Lawyer and writer. Participated in the creation of the Universidade de São Paulo.

Ferreira, Jorge Leal (1928). Physicist. Graduated from the Universidade de São Paulo. Member of the Instituto de Física Teórica.

Ferreira, Paulo Leal (1925). Physicist. Graduated from the Universidade de São Paulo. Founder of the Instituto de Física Teórica.

Ferreira, Ricardo de Carvalho (1928). Professor of chemistry at the Universidade Federal de Pernambuco.

Fonseca, Olímpio Ribeiro da (1895–1978). Parasitologist at the Instituto Oswaldo Cruz, and its director between 1949 and 1953. Worked with the sanitary campaigns of the Rockefeller Foundation in Brazil in the 1910s.

Gama, Lélio (1892–1981). Astronomer and mathematician. Worked in the Observatório do Rio de Janeiro since 1917 and was its director in 1951. Directed the Instituto de Matemática Pura e Aplicada in Rio de Janeiro.

Giesbrecht, Ernesto (1921). Born in Paraná. Studied at the Universidade de São Paulo and was an assistant to Heinrich Rheinboldt. Was vice-director and director of the chemistry department of the Universidade de São Paulo from 1970 to 1978.

Goldemberg, José (1928). Nuclear physicist. Graduated from the Universidade de São Paulo. Worked with Marcelo Damy. Professor at USP. University rector between 1986 and 1989. State secretary for education in São Paulo in 1990, Brazil's secretary for science and technology since March 1990.

Gomes, Francisco Magalhães (1906). Engineer. Born in Ouro Preto and graduated from the Escola de Minas. Became first director of the Instituto de Pesquisas Radioativas in Belo Horizonte in 1954.

Gottlieb, Otto (1920). Researcher in the chemistry of natural products. Worked at the Instituto de Química Agrícola in Rio de Janeiro between 1955 and 1963 and in several universities in Brazil.

Gross, Bernhard (1905). Born in Stuttgart, arrived in Brazil in 1933. Worked at the Instituto Nacional de Tecnologia in Rio de Janeiro and later at the Universidade de São Carlos.

Jacob, Gerhard (1930). Physicist. Born in Hannover, Germany. Studied physics and mathematics at the Universidade do Rio Grande do Sul. In 1956 worked at the Instituto de Energia Atômica of the Universidade de São Paulo. Rector of the Federal University of Rio Grande do Sul since 1988, president of the Conselho Nacional de Desenvolvimento Científico e Tecnológico since March 1990.

Kerr, Warwick Estevam (1922). Agronomist and geneticist. Graduated from the Escola Superior de Agronomia Luiz de Queiroz. Director of the Instituto de Pesquisas da Amazônia from 1975 to 1977.

Lattes, Cesare (1924). Physicist. Student and assistant of Gleb Wataghin's in São Paulo. Worked with Cecil Powell and Giuseppe Occhialini at the University of Bristol in England in 1946. Worked at the Lawrence Radiation Laboratory in Berkeley. Founded the Centro Brasileiro de Pesquisas Físicas in Rio de Janeiro in 1949.

Leinz, Viktor (1904). Geologist. Born in Germany. Came to Brazil in 1935 to work at the Departamento Nacional da Produção Mineral in Rio de Janeiro.

Leite, Rogério Cerqueira (1931). Physicist. Graduated from the Instituto Tecnológico da Aeronáutica. Researcher at the Bell Laboratories. Director of the physics department of the Universidade de Campinas between 1970 and 1975.

Lent, Herman (1911). Medical doctor, helminthologist. Worked at the Oswaldo Cruz Institute from 1932 until going into political exile between 1970 and 1976. Returned to Brazil in 1976.

Leonardos, Othon Henry (1899–1977). Geologist. Graduated from the Escola Politécnica in Rio de Janeiro. Founder and director of the Escola Nacional de Geologia in Rio de Janeiro from 1958 to 1963.

Lopes, Hugo de Souza (1909–1991). Entomologist. Studied with Lauro Travassos. Worked at the Instituto Oswaldo Cruz between 1938 and 1970, when he was forced to leave the military government. Later a professor at the Escola Superior de Agricultura e Medicina Veterinária in Rio de Janeiro.

Lopes, José Leite (1918). Physicist. Studied with Luis Freire and later at the Universidade de São Paulo. Did his doctoral work with Pauli at Princeton. Founder of the Centro Brasileiro de Pesquisas Físicas in Rio de Janeiro. Headed the Department of Nuclear Physics in Strasbourg while in political exile until 1979.

Martins, Amílcar Viana (1907–1990). Parasitologist. Directed the Instituto Nacional de Endemias Rurais between 1956 and 1958. Organized the institute of biological sciences of the Universidade de Minas Gerais.

Mascarenhas, Sérgio. Physicist. Studied with Joaquim Costa Ribeiro in Rio de Janeiro. Works at the Instituto de Física e Química de São Carlos, Universidade de São Paulo.

Mathias, Simão (1908). Chemist. Was an assistant to Heinrich Rheinboldt. Made his career at the Universidade de São Paulo.

Meyer, João Alberto (1925). Born in Poland, arrived in Brazil after World War II. Worked most of his career in Europe, coming to the Universidade de São Paulo and Campinas periodically.

Miller, Harry, Jr. (1895). Parasitologist. born in the United States. Worked for the Rockefeller Foundation in Europe between 1932 and 1941. From 1941 to 1956 coordinated the foundation's medical and natural science program for Latin America.

Mingoia, Quintino (1902). Pharmacologist. Born in Italy, came to Brazil in 1934 to work in a private laboratory. After 1945, taught at the Faculdade de Farmácia e Odontologia of the Universidade de São Paulo.

Miranda, Maury (1930). Molecular biologist, Instituto de Biofísica in Rio de Janeiro.

Monteiro, Hugo Jorge (1938). Pharmacologist. Studied with Carl Djerassi at Stanford University. Worked at the Centro de Pesquisa de Produtos Naturais, Universidade Federal do Rio de Janeiro.

Moreira, Manuel Frota. Physiologist. Worked at the Instituto de Biofísica in Rio de Janeiro. Scientific director of the Conselho Nacional de Pesquisas until 1975.

Mors, Walter Baptista (1920). Chemist. Studied at the Universidade de São Paulo. Worked at the Instituto Agronômico do Norte in 1943 and at the Instituto de Química Agrícola from 1948 to 1964. Professor at the Universidade Federal do Rio de Janeiro.

Paraense, Wladimir Lobato (1914). Biologist. Organized the Serviço Especial de Grandes Endemias of the Instituto Oswaldo Cruz in 1939. Biologist at the institute between 1945 and 1972.

Pavan, Crodovaldo (1919). Geneticist. Was an assistant to René Dreyfus and worked with Dobzhansky in São Paulo and at Columbia University. Was president of the Sociedade Brasileira para o Progresso da Ciência, Fundação de Amparo à Pesquisa de São Paulo, and of the Conselho Nacional de Desenvolvimento Científico e Tecnológico from 1986 to 1990.

Penha, Adolfo Martins (1904–1980). Animal biologist. Worked at the Instituto Biológico de São Paulo.

Pimenta, Aluísio (1923). Pharmacologist. Rector of the Universidade Federal de Minas Gerais from 1964 to 1967.

Pinto, Mário da Silva (1907). Engineer and geologist. Worked at the Departamento Nacional da Produção Mineral in Rio de Janeiro between 1933 and 1951.
Pompéia, Paulus A. (1911). Physicist. Universidade de São Paulo. Organized the physics department of the Instituto Tecnológico da Aeronáutica.
Porto, Sérgio (1926–1977). Physicist. Taught at the Instituto Tecnológico da Aeronáutica. Worked at the Bell Laboratories. Coordinated the research institutes of the Universidade de Campinas.
Reis, José (1907). Bacteriologist. Worked in the Instituto Biológico de São Paulo. Founder of the Sociedade Brasileira para o Progresso da Ciência and editor of *Ciência e Cultura*.
Rezende, Sérgio Machado (1940). Physicist. Ph.D. from Massachusetts Institute of Technology. Founder of the Universidade Federal de São Carlos.
Ribeiro, Darcy (1922). Anthropologist. Graduated from the Escola de Sociologia e Política in São Paulo. Founder and first rector of the Universidade de Brasília.
Sala, Oscar (1922). Physicist. Was an assistant to Gleb Wataghin. Director of the electrostatic accelerator of the Universidade de São Paulo. Presided over the Sociedade Brasileira para o Progresso da Ciência.
Salmeron, Roberto. Physicist. Graduated in engineering from the Escola Politécnica in São Paulo. Coordinator of the scientific institutes of the Universidade de Brasília between 1961 and 1965.
Salzano, Francisco Mauro (1928). Geneticist. Studied with Antônio Cordeiro in Rio Grande do Sul. Later worked with Dobzhansky and Pavan.
Sawaya, Paulo (1904). Zoologist. Studied with Alfonso Bovero. Made his career at the Universidade de São Paulo.
Schenberg, Mário (1914–1990). Physicist. Studied with Gleb Wataghin and worked with Fermi, Pauli, and Gamov. Director of the physics department of the Universidade de São Paulo between 1953 and 1961.
Senise, Pascoal Américo (1917). Graduated in chemistry from the Universidade de São Paulo in 1942. Director of the Instituto de Química of that university until 1970.
Silva, Maurício Rocha e (1910–1983). Pharmacologist. Worked in the Instituto Biológico from 1943 to 1957. Founder of the Sociedade Brasileira para o Progresso da Ciência.
Tiomno, Jayme (1920). Physicist. Studied with Gleb Wataghin and Mario Schenberg at the Universidade de São Paulo and later at Princeton. Worked at the Universidade de Brasília and the Universidade Católica do Rio de Janeiro.
Vale, José Ribeiro do. Pharmacologist. Graduated from the Faculdade de Medicina in São Paulo. Worked at the Escola Paulista de Medicina until retirement, and organized and directed its department of biochemistry and pharmacology.
Vanzolini, Paulo Emílio (1923). Biologist. Ph.D. from Harvard University. Director of the Museu de Zoologia of the Universidade de São Paulo.
Vargas, José Israel (1928). Physicist. Studied at the Universidade de São Paulo. Did his doctoral studies in Cambridge. Worked between 1965 and 1972 at the University of Grenoble and afterward at the Universidade Federal de Minas Gerais.
Vaz, Zeferino (1908–1981). Parasitologist. Studied with Lauro Travassos.

Founded and directed the Escola de Medicina de Ribeirão Preto and the Universidade de Campinas.

Wladislaw, Blanka. Born in Poland. Graduated in chemistry from the Universidade de São Paulo. In 1971 became head of the department of fundamental chemistry at the institute of chemistry of the Universidade de São Paulo.

Other Interviews

Beviláqua, Luís, and Giulo Massarani, Sérgio Neves Monteiro, Carlos Augusto Perlingeiro, and Luís Pinguelli Rosa. Engineers. Interviewed by Márcia B. de Melo Nunes and Nadja X. Souza for the study on the engineering program of the Universidade Federal do Rio de Janeiro (COPPE), 1978 (Nunes, Souza & Schwartzman 1982).

Montenegro, Casimiro. Brigadier General. Interviewed for the Centro de Pesquisa e Documentação em História Contemporânea do Brasil da Fundação Getúlio Vargas (CPDOC) by Simon Schwartzman and Cláudia Guimarães, with the participation of José Pelúcio Ferreira, José Israel Vargas, and Técio Pacciti, 1988.

Peixoto, Maurício Mattos, and Elon Lages de Lima. Mathematicians. Interviewed by João Batista de Oliveira's project on scientific careers (J. B. A. Oliveira 1984).

Wataghin, Gleb. Physicist. Interviewed by Cylon T. Gonçalves, Physics Institute, Universidade de Campinas, 1976.

BIBLIOGRAPHICAL REFERENCES

Abreu, Sílvio F. 1975. *Riqueza mineral do Brasil.* 2nd ed. São Paulo: Cia. Editora Nacional.
Adler, Emanuel. 1987. *The Power of Ideology: The Quest for Technological Autonomy in Argentina and Brazil.* Berkeley and Los Angeles: University of California Press.
Adorno, Sérgio. 1988. *Os aprendizes do poder—o bacharelismo liberal na política brasileira.* Rio de Janeiro: Paz e Terra.
Albagli, S. 1987. "Marcos institucionais do Conselho Nacional de Pesquisas." *Perspicillum* (Rio de Janeiro, Museu de Astronomia e Ciências Afins), 1 (May): 1–166.
Albernaz, Paulo M. 1968. *A Escola Paulista de Medicina* (*Notícia histórica dos primeiros vinte e cinco anos*). São Paulo: Escola Paulista de Medicina.
Albertin, P., and T. Faria. 1984. "Arte e ciência no Brasil holandês." *Ciência Hoje* (Rio de Janeiro), 3 (November–December): 34–41.
Alden, Dauril. 1968. *Royal Government in Colonial Brazil.* Berkeley and Los Angeles: University of California Press.
Alden, Dauril, ed. 1972. *Colonial Roots of Modern Brazil.* Berkeley and Los Angeles: University of California Press.

Almeida, A., Jr. 1956. *Problemas do ensino superior.* São Paulo: Cia. Editora Nacional.
Alves, Márcio M. 1979. *A Igreja e a Política no Brasil.* São Paulo: Editora Brasiliense.
Amaral, Afrânio do. 1958. "Evolução dos institutos científicos." In Anhembi 1958:376–96.
Anhembi. 1958. *Ensaios Paulistas.* São Paulo: Editora Anhembi.
Azevedo, Fernando de. 1958. "A Universidade de São Paulo." In Anhembi 1958:215–26.
———. 1963. *A cultura brasileira.* 4th ed. (1st ed. 1940). Brasília: Editora da Universidade de Brasília.
Azevedo, Fernando de, ed. 1955. *As ciências no Brasil.* 2 vols. São Paulo: Editora Melhoramentos.
———. 1971. *História da minha vida.* Rio de Janeiro: Editora José Olímpio.
Azevedo, Moreira de. 1885. "Sociedades fundadas no Brasil desde os tempos coloniais até o começo do segundo reinado." *Revista do Instituto Histórico e Geográfico do Brasil* 68, no. 2:411–18.
Bacha, Edmar, and Herbert S. Klein, eds. 1986. *A transição incompleta. Brasil desde 1945.* 2 vols. Rio de Janeiro: Paz e Terra.
Balán, Jorge. 1973. "Migrações e desenvolvimento capitalista no Brasil." *Estudos CEBRAP* 5:5–80.
Barata, Mário. 1973. *Escola Politécnica do Largo de São Francisco. Berço da engenharia brasileira.* Rio de Janeiro: Associação dos Antigos Alunos da Escola Politécnica e Clube de Engenharia.
Barbosa, Plácido, and C. Barbosa Rezende. 1909. *Os serviços de saúde pública no Brasil, especialmente na cidade do Rio de Janeiro, de 1880 a 1907 (Histórico de legislação).* Rio de Janeiro: Diretoria Geral de Saúde Pública, Imprensa Nacional.
Barnes, B. 1974. *Scientific Logic and Sociological Theory.* London: Routledge & Kegan Paul.
Barros, R. S. Maciel de. 1959. *A ilustração brasileira e a idéia de universidade.* São Paulo: Universidade de São Paulo, Faculdade de Filosofia, Ciências e Letras. A new edition is available from Convívio and EDUSP, 1987.
———. 1962. "Vida religiosa." In Holanda 1962, 4:317–37.
Basalla, George. 1967. "The Spread of Western Science." *Science* 156, no. 3775:611–22.
Bastide, Roger. 1951. "Religion and Church in Brazil." In Smith and Marchant (eds.) 1951:334–55.
Bauer, Peter. 1977. "Reflections on Western Technology and 'Third World' Development." *Minerva* 15 (Summer): 144–54.
Bella, Robert N. 1971. *Continuity and Change in Japanese Society.* Boston: Little, Brown.
Benchimol, Jaime L., ed. 1990. *Manguinhos do sonho à vida—A ciência na Belle Époque.* Rio de Janeiro: Casa de Oswaldo Cruz, Fiocruz.
Ben-David, Joseph. 1971. *The Scientist's Role in Society: A Comparative Study.* Englewood Cliffs, N.J.: Prentice-Hall.
———. 1976. "Report on Visit to Brazil." Rio de Janeiro: FINEP. Ms. Published in Portuguese as "Universidade e ciência observadas por Ben-David. Relatório de uma visita ao Brasil (25 de julho–8 de agosto de 1976)." *Ciência Hoje* 7 (November 1987): 68–73.

———. 1977. *Centers of Learning: Britain, France, Germany, the United States.* Berkeley, Calif.: Carnegie Commission on Higher Education.
Bendix, Reinhard. 1984. *Force, Fate & Freedom: On Historical Sociology.* Berkeley and Los Angeles: University of California Press.
Berlink, Ciro, and A. Trujillo Ferrari. 1958. *A Escola de Sociologia e Política de São Paulo, 1933–1958.* São Paulo: Escola de Sociologia e Política.
Bernal, J. D. 1971. *Science in History.* 4 vols. Cambridge: MIT Press.
Bloor, David. 1976. *Knowledge and Social Imagery.* London: Routledge & Kegan Paul.
Blount, J. A. 1971. "The Public Health Movement in São Paulo, Brazil: A History of the Sanitary Service, 1892–1918." Ph.D. diss., Tulane University, Department of History.
BNDE (Banco Nacional de Desenvolvimento Econômico). 1974. *FUNTEC—10 anos de apoio à pesquisa.* Rio de Janeiro: BNDE.
Botelho, Antônio José. 1983. "Les Scientifiques et le pouvoir au Brésil: Le Cas de la Société Brésilienne por le Progrès de la Science (SBPC), 1948–1980." Paris: Conservatoire National des Arts et Métiers, Centre de Science, Technologie et Société. Mimeograph.
Boxer, C. R. 1965. *The Dutch in Brazil, 1624–1654.* Hamden, Conn.: Shoe String Press.
———. 1973. *The Dutch Seaborne Empire, 1600–1800.* New York: Knopf.
Braga, Ernani. 1984. *O pensamento de Ernani Braga.* Edited by Paulo Marchiori Buss. Rio de Janeiro, Escola Nacional de Saúde Pública.
Brasil. Câmara dos Deputados, Comissão de Ciência e Tecnologia. 1973. *Ciência, tecnologia e desenvolvimento.* Brasília: Câmara dos Deputados.
Brasil. Conselho Nacional de Desenvolvimento Científico e Tecnológico. 1982. *Orçamento da União para ciência e tecnologia. Anotações e destaques.* Brasília: CNPq, Coordenação Editorial.
Bruneau, Thomas C. 1974. *The Political Transformations of the Brazilian Catholic Church.* New York: Cambridge University Press.
Burtt, Edwin A. 1951. *The Metaphysical Foundations of Modern Physical Science.* New York and London: Humanities Press and Routledge & Kegan Paul.
Campos, Ernesto de Souza. 1941. *Instituições culturais e educação superior no Brasil. Resumo histórico.* Rio de Janeiro: Imprensa Nacional.
———. 1954. *História da Universidade de São Paulo.* São Paulo: Universidade de São Paulo.
Campos, Francisco. 1940. *O Estado nacional. Sua estrutura, seu conteúdo ideológico.* Rio de Janeiro: José Olímpio.
Cardwell, D. S. L. 1972. *The Organization of Science in England.* London: Heinemann Educational Books.
Carneiro, J. F. D., et al. 1969. *Relatório da equipe de assessoria ao planejamento do ensino superior (Acordo Ministério da Educação Cultura–United States Agency for International Development).* Rio de Janeiro: Ministério da Educação e Cultura.
Carone, Edgar. 1971. *A República Velha.* São Paulo: Difel.
Cartaxo, Ernani. 1948. "Histórico da Universidade do Paraná." In *Anuário da Universidade do Paraná, 1946–1947* Curitiba: Universidade do Paraná.
Carvalho, Hervásio G. 1973. "Pesquisa básica e desenvolvimento nuclear." In *Brasil 1973*:115–52.

Carvalho, J. de. 1950. *Subsídios para a história da filosofia e da ciência em Portugal*. 2 vols. Coimbra: Universidade de Coimbra.
Carvalho, J. Murilo. 1978. *A Escola de Minas de Ouro Preto. O peso da glória*. Rio de Janeiro and São Paulo: Cia. Editora Nacional and FINEP.
———. 1980. *A construção da ordem*. Rio de Janeiro: Editora Campus.
———. 1987. *Os bestializados*. São Paulo: Companhia das Letras.
Castro, Antônio de Barros. 1971. *Sete ensaios sobre a economia brasileira*, 2 vols. Rio de Janeiro. Editora Forense.
Castro, Antônio de Barros, and Pires Souza. 1985. *A economia brasileira em ritmo de marcha forçada*. Rio de Janeiro: Paz e Terra.
Castro, Cláudio de Moura. 1986a. "Há produção científica no Brasil?" In Schwartzman and Castro 1986:190–224.
———. 1986b. "O que está acontecendo com a educação no Brasil?" In Bacha and Klein 1986, 2:103–62.
Castro, Cláudio de Moura, and Gláucio A. Soares. 1986. "As avaliações da CAPES." In Schwartzman and Castro 1986: 190–224.
Castro, F. M. de Oliveira. 1955. "A matemática no Brasil." In F. de Azevedo 1955, 1:41–77.
Cava, R. Della. 1976. "Catholicism and Society in Twentieth Century Brazil." *Latin American Research Review* 12, no. 2: 7–50.
Chacel, Julien M., Pamela S. Falk, and David V. Fleischer, eds. 1988. *Brazil's Economic and Political Future*. Boulder, Colo.: Westview Press.
Chur, L. A., D. E. Bertels, B. Komissarov, and N. Licenko. 1981. *A expedição científica de G. I. Langsdorff ao Brasil, 1821–1829*. Translated from Russian by Marcos Pinto Braga. Brasília: Secretaria do Patrimônio Histórico e Artístico Nacional, Fundação Nacional Pró-Memória.
Cidade, Hernani. 1969. "A reforma pombalina da instrução." In *Lições de cultura e literatura portuguesa*. 2 vols. Coimbra: Coimbra Editora.
Clark, Burton, R., ed. 1984. *Perspectives in Higher Education: Eight Disciplinary and Comparative Views*. Berkeley and Los Angeles: University of California Press.
Conniff, Michael. 1981. *Urban Politics in Brazil: The Rise of Populism, 1925–1945*. Pittsburgh, Pa.: University of Pittsburgh Press.
Costa, Amoroso. 1971. *Idéias fundamentais da matemática e outros ensaios*. São Paulo: Grijalbo and Editora da Universidade de São Paulo.
CPDOC (Centro de Pesquisa e Documentação em História Contemporânea do Brasil). 1984. *História da ciência no Brasil (Acervo de depoimentos)*. Rio de Janeiro: CPDOC and FINEP.
Crosland, Maurice, ed. 1976. *The Emergence of Science in Western Europe*. New York: Science History Publications.
Cruz, Isar Hasselman Oswaldo. 1985. Letter to *Ciência Hoje* 3 (January–February): 5.
Dahlman, Carl J., and Fernando Valadares Fonseca. 1987. "From Technological Dependency to Technological Development: The Case of Usiminas Steelplant in Brazil." In Katz (ed.) 1987:154–82.
Daland, Robert T. 1967. *Brazilian Planning: Development Politics and Administration*. Chapel Hill: University of North Carolina Press.
D'Alessandro, Alexandre. 1943. *A Escola Politécnica de São Paulo. A história de sua história*. 3 vols. São Paulo: Revista dos Tribunais.
Dean, Warren. 1989. "The Green Wave of Coffee: Beginnings of Tropical Agri-

cultural Research in Brazil (1885–1900)." *Hispanic American Historical Review* 69, no. 1: 91–116.
Dedijer, Stevan. 1963. "Underdeveloped Science and Underdeveloped Countries." *Minerva* 2:61–87.
Delfim Neto, A. 1959. *O problema do café no Brasil.* São Paulo: Faculdade de Ciências Econômicas e Administrativas da Universidade de São Paulo.
Domingues, Mário. 1963. *O Marquês de Pombal e sua época.* 2nd ed. Lisbon: Editora Romano Torres.
Duarte, Paulo. 1976. *Memórias. Selva obscura.* Vol. 3. São Paulo: Editora Hucitec.
Eisenstadt, S. N., and S. Rokkan. eds. 1973. *Building States and Nations.* 2 vols. Beverly Hills, Calif.: Sage Publications.
Erber, Fábio. 1977. "Technological Development and State Intervention: A Study of the Brazilian Capital Goods Industry." Ph.D. diss., University of Sussex.
Evans, Peter B. 1986. "State, Capital, and the Transformation of Dependence: The Brazilian Computer Case." *World Development* 14, no. 7:791–808.
Falcão, Edgard Cerqueira, ed. 1965. *Obras científicas, políticas e sociais de José Bonifácio de Andrada e Silva.* São Paulo: private printing.
———. 1973. *Oswaldo Cruz Monumenta Historica.* 3 vols. São Paulo: private printing.
———. 1974. *Gazeta Médica da Bahia* (facsimile). 3 vols. São Paulo: private printing.
Falcon, Francisco José Calazans. 1982. *A época pombalina. Política econômica e monarquia ilustrada.* São Paulo: Editora Ática.
Faoro, Raymundo. 1958. *Os donos do poder. Formação do patronato político brasileiro.* Porto Alegre: Editora Globo.
Faria, L. de Castro. 1949. *As exposições de antropologia e arqueologia do Museu Nacional.* Rio de Janeiro: Imprensa Nacional.
———. 1951. "J. B. Lacerda e as pesquisas antropológicas no Brasil." Rio de Janeiro, Museu Nacional, publicações avulsas no. 6.
Faria, Vilmar. 1986. "Mudanças na composição do emprego e na estrutura das ocupações." In Bacha and Klein (eds.) 1986, 1:75–112.
Fausto, Bóris, ed., *História geral da civilização brasileira.* Part 3: *O Brasil republicano.* São Paulo: Difel, 1975.
Ferri, Mário G. 1955. "A botânica no Brasil." In F. de Azevedo (ed.) 1955, 2:149:200.
———. 1980. "História da botânica no Brasil." In Ferri and Motoyama (eds.) 1979–81, 2:33–88.
Ferri, Mário G., and Shozo Motoyama, eds. 1979–81. *História da ciência no Brasil.* 3 vols. São Paulo: Editora da Universidade de São Paulo.
FIBGE (Fundação Instituto Brasileiro de Geografia e Estatística). 1987. *Anuário estatístico do Brasil, 1986.* Rio de Janeiro: FIBGE.
Fonseca Filho, O. 1974. "A Escola de Manguinhos." In Falcão (ed.) 1973, 2:11–300.
Franca, Leonel. 1952. *O método pedagógico dos jesuítas.* Rio de Janeiro: Editora Agir.
Frieiro, Eduardo. 1982. *O diabo na livraria do cônego.* São Paulo: Editora Itatiaia and Editora da Universidade de São Paulo.
Frischtak, Cláudio. 1986. "Brazil." In Rushing and Brown 1986:31–70.
Furtado, Celso. 1968. *The Economic Growth of Brazil: A Survey from Colonial to Modern Times.* Berkeley and Los Angeles: University of California Press.

Furtado, Jacundino. 1962. *Universidade do Paraná, 1912–1962. Publicação do seu cinquentenário.* Curitiba: Universidade Federal do Paraná.
Gall, Norman. 1976. "Atoms for Brazil, Dangers for All?" *Foreign Policy* 23 (Summer): 155–201.
Gama, Lélio. 1971. "A obra de Amoroso Costa." In Costa 1971:27–37.
Garfield, E. 1983. "Mapping Science in the Third World." *Science and Public Policy* 10 (June): 112–27.
Gerth, H. H., and C. H. Mills, eds. 1958. *From Max Weber: Essays in Sociology.* New York: Oxford University Press.
Gibbons, M., and B. Wittrock, eds. 1985. *Science as a Commodity: Threats to the Open Community of Scholars.* Essex: Longman Group.
Giddens, Antony. 1979. *Central Problems in Social Theory.* London: Macmillan.
———. 1987. *Social Theory and Modern Sociology.* Stanford, Calif.: Stanford University Press.
Gilpin, Robert. 1968. *France in the Age of the Scientific State.* Princeton, N.J.: Princeton University Press.
Glaser, William A. 1978. *The Brain Drain: Migration and Return.* Oxford: Pergamon Press and UNITAR.
Godinho, V. M. 1961–70. "Portugal and Her Empire." In *Cambridge Modern History.* Vols. 5 and 6. New York: Cambridge University Press.
Gould, S. J. 1977. *Ever Since Darwin: Reflections in Natural History.* New York: W. W. Norton.
Graciarena, Jorge. 1964. *Formación de postgrado en ciencias sociales en América Latina.* Buenos Aires: Editorial Paidós.
Graham, Douglas H. 1973. "Migração estrangeira e a questão da oferta de mão de obra no crescimento econômico brasileiro, 1880–1930." *Estudos Econômicos* 3, no. 2: 7–64.
Graham, Richard. 1968. *Britain and the Onset of Modernization in Brazil, 1850–1914.* Cambridge: Cambridge University Press.
Guerra, E. Sales. 1940. *Oswaldo Cruz.* Rio de Janeiro: Editora Vecchi.
Guilherme, Olímpio. 1957. *O Brasil na era atômica.* Rio de Janeiro: Editora Vitória.
Hashimoto, U. 1963. "An Historical Synopsis of Education and Science in Japan from the Meiji Restoration to the Present Day." *The Impact of Science in Society* 13, no. 1: 3–23.
Herrera, Amílcar. 1971. *Ciencia y política en América Latina.* Mexico: Siglo XXI.
Holanda, Sérgio Buarque. 1960a. "Franceses, ingleses e holandeses no Brasil quinhentista." In Holanda 1960b, 1:147–234.
———. 1960b. *História geral da civilização brasileira.* Part 1: *A época colonial*, 2 vols. São Paulo: Difusão Européia do Livro. 2nd ed. 1975.
———. 1962. *História geral da civilização brasileira.* Part 2: *O Brasil monárquico*, 5 vols. São Paulo: Difusão Européia do Livro. 2nd ed. 1976–78.
Höning, Chaim S., and Elza F. Gomide. 1979. "Ciências matemáticas." In Ferri and Motoyama 1979–81, 1:35–60.
Instituto Agronômico de Campinas. 1977. *Instituto Agronômico. Histórico, organização, atividades, 1887–1977.* São Paulo: Imprensa Oficial.
Instituto de Biofísica da Universidade do Rio de Janeiro. 1951. *Homenagem a Guilherme Guinle.* Rio de Janeiro: Universidade do Rio de Janeiro.
Instituto de Tecnologia Industrial. 1958. "Histórico e atuação do Instituto de Tecnologia Industrial no desenvolvimento técnico científico em Minas

Gerais." *Boletim do Instituto de Tecnologia Industrial* (Belo Horizonte), 25:1–42.
Jornal do Brasil. 1987. "Brasil domina a tecnologia do combustível nuclear." *Jornal do Brasil*, September 5, p. 5.
Katz, Jorge, ed. 1987. *Technology Generation in Latin American Manufacturing Industries*. New York: St. Martin's Press.
Keith, H., and S. F. Edwards. 1969. *Conflict and Continuity in Brazilian Society*. Columbia: University of South Carolina Press.
Knorr, K., and R. Whitley. 1981. *Sociology of the Sciences: The Social Process of Scientific Investigation*. Boston: D. Reidel.
Knorr-Cetina, K. 1981. *The Manufacture of Knowledge: An Essay on the Constructivist and Contextual Nature of Science*. Oxford: Pergamon Press.
Knorr-Cetina, K., and M. Mulkay. 1983. *Science Observed*. Beverly Hills, Calif.: Sage Publications.
Koizumi, K. 1975. "The Emergence of Japan's First Physicists, 1868–1900." *Historical Studies in the Physical Sciences*, 6: 3–108.
Kuhn, Thomas S. 1970. *The Structure of Scientific Revolutions*. 2nd ed. Chicago: University of Chicago Press.
———. 1977. *The Essential Tension: Selected Studies in Scientific Tradition and Change*. Chicago: University of Chicago Press.
Laboriau, Ferdinand, Roquete Pinto, and Licínio Cardoso, eds. 1929. *O problema universitário brasileiro (Inquérito promovido pela seção de ensino técnico e superior da Associação Brasileira de Educação)*. Rio de Janeiro: Editora "A Encadernadora."
Lacaz, Carlos da Silva. 1977. *Vultos da medicina brasileira*. 4 vols. São Paulo: Laboratórios Pfizer do Brasil.
Lacerda, João Batista de. 1905. *Fatos do Museu Nacional do Rio de Janeiro (Recordações históricas e científicas fundadas em documentos autênticos e informações verídicas)*. Rio de Janeiro: Imprensa Nacional.
Lacombe, Américo Jacobina. 1960. "A igreja no Brasil colonial." In Holanda 1960b, 1:51–57.
Ladosky, Waldemar. 1985. Letter to *Ciência Hoje* 3 (May–June): 4–5.
Lang, James. 1979. *Portuguese Brazil: The King's Plantation*. New York: Academic Press.
Latour, B. 1987. *Science in Action*. Cambridge: Harvard University Press.
Latour, B., and S. Woolgar. 1979. *Laboratory Life: The Social Construction of Scientific Facts*. Beverly Hills, Calif.: Sage Publications.
Leff, Nathanael H. 1968. *Economic Policy-making and Development in Brazil, 1947–1964*. New York: John Wiley & Sons.
Leinz, Viktor. 1955. "A geologia e a paleontologia no Brasil." In F. de Azevedo 1955, 1:243–63.
Leite, Rogério C. C. 1977. *Energia nuclear e outras mitologias*. São Paulo: Duas Cidades.
Leonardos, Othon. 1955. "A mineralogia e a petrografia no Brasil." In F. de Azevedo 1955, 1:267–33.
Lessa, Carlos. 1978. "A estratégia de desenvolvimento, 1974–1978. Sonho e fracasso." Rio de Janeiro: Universidade Federal do Rio de Janeiro, Faculdade de Economia e Administração. Mimeograph.
Lévi-Strauss, Claude. 1955. *Tristes Tropiques*. Paris: Plon.
Levy, Daniel C. 1986. *Higher Education and the State in Latin America: Private Challenges to Public Dominance*. Chicago: University of Chicago Press.

Lins, Ivan M. de Barros. 1967. *História do positivismo no Brasil*. São Paulo: Cia. Editora Nacional.
Lobato, Monteiro. 1936. *O escândalo do petróleo*. São Paulo: Cia. Editora Nacional.
Lobo, Francisco Bruno. 1964-1969. *O ensino da medicina no Rio de Janeiro*. Rio de Janeiro: Revista do Instituto Histórico e Geográfico.
———. 1969. *Uma universidade no Rio de Janeiro*. Rio de Janeiro: Ministério da Educação (CAPES) and Universidade Federal do Rio de Janeiro.
Lopes, José Leite. 1978. *Ciência e libertação*. Rio de Janeiro: Paz e Terra.
———. 1988. *Richard Feynman e a física no Brasil*. Rio de Janeiro: Centro Brasileiro de Pesquisas Físicas. Publicação CBPC-CS-005/88.
Machado, Roberto, Angela Loureiro, Rogério Luz, and Kátia Muricy. 1978. *Danação da norma. Medicina social e constituição da psiquiatria no Brasil*. Rio de Janeiro: Graal.
Magalhães, C. 1967. *História do pensamento econômico em Portugal*. Coimbra: Editora Coimbra.
Magalhães, Fernando. 1932. *O centenário da Faculdade de Medicina do Rio de Janeiro, 1832–1932*. Rio de Janeiro: Tipografia A. B. Bartthel.
Manchester, Alan K. 1933. *British Preeminence in Brazil, Its Rise and Decline*. Chapel Hill: University of North Carolina Press.
———. 1969. "The Transfer of the Portuguese Court to Rio de Janeiro." In Keith and Edwards 1969:148–83.
Marchant, Alexander. 1961. "Aspects of the Enlightenment in Brazil." In Whitaker 1961:95–118.
Marchant, Anyda. 1961. "D. José Botanical Garden." *Hispanic American Historical Review* 41, no. 2: 259–74.
Mariani, M. Clara. 1982a. "Educação e ciências sociais. O Instituto Nacional de Estudos e Pesquisas Educacionais." In Schwartzman 1982:169–95.
———. 1982b. "O Instituto de Biofísica da Universidade Federal do Rio de Janeiro." In Schwartzman 1982:199–208.
Martins, Tales. 1955. "A biofísica no Brasil (Episódios de sua história)." In F. de Azevedo 1955, 2:201–59.
Martins Filho, Amílcar, and Roberto B. Martins. 1983. "Slavery in a Nonexport Economy: Nineteenth-Century Minas Gerais Revisited." *Hispanic American Historical Review* 63, no. 3:537–68.
Mason, S. F. 1975. *A History of Sciences*. New York: Collier Books.
Mathias, Simão. 1975. *Cem anos de química no Brasil*. São Paulo: Universidade de São Paulo. Coleção Revista de História.
Maxwell, K. 1972. "The Generation of the 1870s and the Idea of the Luso-Brazilian Empire." In Alden 1972: 107–46.
McCann, Frank D., Jr. 1973. *The Brazilian-American Alliance, 1937–1945*. Princeton, N.J.: Princeton University Press.
McLeod, R. M. 1975. "Scientific Advice for British India: Imperial Perceptions and Administrative Goals, 1898–1923." *Modern Asia Studies* 9, no. 3: 343–84.
Meiller, J. Luís, and Francisco I. A. Silva. 1949. "Meio século de tecnologia, 1899–1949." *Boletim do Instituto de Pesquisas Tecnológicas* (São Paulo), 34.
Melo, J. A. G. 1976. "O domínio holandês na Bahia e no Nordeste." In Holanda 1962, 1:235–53.
Merton, Robert K. 1938. "Science and the Social Order." In Merton 1973, chap. 12.

———. 1957. *Social Theory and Social Structure*. Glencoe, Ill.: Free Press.
———. 1970. *Science, Technology, and Society in Seventeenth-Century England*. New York: Harper & Row.
———. 1973. *The Sociology of Science: Theoretical and Empirical Investigations*. Ed. Norman W. Storer. Chicago: University of Chicago Press.
Mesquita Filho, J. 1969. *Política e cultura*. São Paulo: Livraria Martins.
Miceli, S. 1990. *História das Ciências Sociais no Brasil*. São Paulo: Ed. Vértice, 1989.
Ministério de Ciência e Tecnologia. 1985. *Ministério de Ciência e Tecnologia—Ano 1 (Relatório de atividades)*. Brasília: MCT.
Morais, Abraão. 1955. "A astronomia no Brasil." In F. de Azevedo 1955, 1:81–161.
Moravcsik, Michael J. 1975. *Science Development*. Bloomington, Ind.: Pasitam.
Mora y Araujo, M., ed. 1983. *Política tecnológica y países en desarrollo*. Buenos Aires: Editorial del Instituto.
Morehause, W. 1971. *Science in India: Institution Building and the Organizational System in Historical Perspective*. Bombay: College of India and Popular Prakastan.
Morel, Regina L. Morais. 1979. *Ciência e estado. A política científica no Brasil*. São Paulo: T. A. Queirós.
Morel, Regina L. Morais, and Carlos Morel. 1977. "Um estudo sobre a produção científica brasileira segundo os dados do I.S.I." *Ciência da Informação* 6, no. 2: 99–109.
Morize, H. 1987. *Observatório Astronômico. Um século de história (1827–1027)*. Rio de Janeiro: Museu de Astronomia e Ciências Afins.
Mörner, M., ed. 1965. *The Expulsion of the Jesuits from Latin America*. New York: Knopf.
Mors, Walter B. 1985. "A árvore da ciência." Letter to *Ciência Hoje* 3 (January–February); 5.
Mota, Ivone Freire, and Amélia Império Hamburger. 1988. "Retratos de Luís de Barros Freire como pioneiro da ciência no Brasil." *Ciência e Cultura* 40 (September): 875–81.
Moyal, A. M. 1976. *Scientists in Nineteenth-Century Australia: A Documentary History*. Melbourne: Cassel Australia.
Mulkay, M. J. 1977. "Sociology of the Scientific Research Community." In Spiegel-Rösing and Price 1977: 93–196.
Museu de Astronomia e Ciências Afins. 1988. *Arquivo Lélio Gama. Inventário sumário*. Rio de Janeiro: MAST.
Museu Nacional. 1951. *Comemoração do centenário de J. B. Lacerda, 1846–1946*. Rio de Janeiro: Museo Nacional, Avulso no. 6.
Museu Paraense Emílio Goeldi. 1986. *O Museu Paraense Emílio Goeldi*. São Paulo: Banco Safra.
Nachman, R. G. 1977. "Positivism and Revolution in Brazil's First Republic: The 1904 Revolt." *The Americas* 34, no. 1: 20–39.
Nau, Henry. 1986. "National Policies for High Technology Development and Trade: An International Comparative Assessment." In Rushing and Brown 1986:9–30.
Needell, Jeffrey D. 1987. "The *Sublime Porte:* French Influence on Brazilian Literature and Literati, 1808–1914." Washington, D.C.: Woodrow Wilson International Center for Scholars. Mimeograph.
Neiva, Artur. 1941. "Adolfo Lutz." *Memórias do Instituto Oswaldo Cruz* 36, no. 1:i–ix.

Nielsen, Waldemar A. 1972. *The Big Foundations.* New York: Columbia University Press.
Novais, Fernando A. 1981. *Portugal e Brasil na crise do antigo sistema colonial, 1777– 1808.* São Paulo: Editora Hucitec.
Nunes, Márcia Bandeira de Melo, Nadja V. X. Souza, and S. Schwartzman. 1982. "Pós-graduação em engenharia. A experiência da COPPE." In Schwartzman (ed.) 1982:209–43.
Oberakcker, Carlos. 1960. "Viajantes, naturalistas e artistas estrangeiros." In Holanda 1962, 1:119–34.
Oliveira, João Batista A. 1984. *Ilhas de competência. Carreiras científicas no Brasil.* São Paulo: Editora Brasiliense.
Oliveira, Lúcia Lippi. 1986. "Donald Pierson e a sociologia no Brasil." Paper presented to the 10th Annual Meeting of the Brazilian Association for Graduate Programs in the Social Sciences (ANPOCS), Campos de Jordão.
Oliveira, Neide S. 1975. "Cientista. O indivíduo e a ocupação." M.A. thesis, Universidade de São Paulo, Faculdade de Filosofia, Ciências e Letras. Mimeograph.
Paim, Antônio. 1971. *O conceito de ciência na obra de José Bonifácio (Textos escolhidos).* Rio de Janeiro: Universidade Católica, Departamento de Filosofia.
———. 1974. *História das idéias filosóficas no Brasil.* São Paulo: Grijalbo and Editora da Universidade de São Paulo.
———. 1982. "Por uma universidade no Rio de Janeiro." In Schwartzman 1982:17–96.
———. 1987. *O modelo de desenvolvimento tecnológico implantado pela Aeronáutica.* Rio de Janeiro: Centro de Comunicação Social da Aeronáutica.
Paulinyi, Erno, et al. 1986. "Indicadores básicos de ciência e tecnologia." Brasília: Conselho Nacional de Desenvolvimento Científico e Tecnológico. Mimeograph.
Pena, M. Valéria. 1977. "A evolução da pesquisa de saúde no Brasil. Uma interpretação preliminar." Rio de Janeiro: Conselho Nacional de Desenvolvimento Científico e Tecnológico. Mimeograph.
Penna, Maria Luiza. 1987. *Fernando de Azevedo. Educação e transformação.* São Paulo: Editora Perspectiva.
Pereira, Jesus Soares. 1975. *Petróleo, energia elétrica, siderurgia. A luta pela emancipação.* Text edited and commented on by Medeiros Lima. Rio de Janeiro: Paz e Terra.
Pereira, J. Veríssimo da Costa. 1955. "A Geografia no Brasil." In F. de Azevedo 1955, 1:317–412.
Pereira, Vera Maria C. 1978. "Cooperação internacional para a ciência e tecnologia no Brasil." Rio de Janeiro: FINEP.
Picaluga, I., A. C. Torres Ribeiro, and N. Costa. 1977. "Campanhas sanitárias e a institucionalização da saúde pública no Brasil." Rio de Janeiro: FINEP. Mimeograph.
Pimenta, Aluísio. 1984. *Universidade. A destruição de uma experiência democrática.* Rio de Janeiro: Editora Vozes.
Pinto, Mário da Silva. 1985. Letter to *Ciência Hoje* 3 (March–April): 2.
Pinto, O. M. de Oliveira. 1955. "A zoologia no Brasil." In Azevedo 1955, 2:93–148.
Pinto, Ricardo G. Ferreira. 1978. "Liliputianos e lapucianos. Os caminhos da

física no Brasil (1810 a 1949)." M.A. thesis, Instituto Universitário de Pesquisas do Rio de Janeiro. Mimeograph.
Piragibe, Clélia. 1985. *Indústria de informática. Desenvolvimento brasileiro e mundial.* Rio de Janeiro: Editora Campus.
Polanyi, Michael. 1962. *Personal Knowledge: Towards a Post-Critical Philosophy.* London: Routledge & Kegan Paul.
———. 1968. "The Republic of Science, Its Political and Economic Theory." In Shils 1968:1–21.
Porto, Ângela. 1987. "Positivismo e seus dilemas." *Ciência Hoje* 6 (August): 54–61.
Prado, Antônio de Almeida. 1958. "Quatro séculos de medicina na cidade de São Paulo." In Anhembi 1958:769–802.
Prado, Caio, Jr. 1967. *The Colonial Background of Modern Brazil.* Berkeley and Los Angeles: University of California Press.
Prado, Leal. 1975. "Evolução da bioquímica no Brasil." *O Estado de São Paulo.* Suplemento do centenário, no. 15.
Price, D. J. Solla. 1963. *Little Science, Big Science.* New York: Columbia University Press.
Pyenson, Lewis. 1982. "Cultural Imperialism and Exact Sciences: German Expansion Overseas, 1900–1930." *History of Science* (London), 20:1–43.
———. 1984. "*In Partibus Infidelum:* Imperialist Rivalries and Exact Sciences in the Early Twentieth Century in Argentina." *Quipu* 1, no. 2: 253–303.
Rahman, A. 1970. "Scientists in India: The Impact of Economic Policies and Support in Historical Perspective." *International Social Sciences Journal* 22(1): 59–88.
Reis, Elisa P. 1979. "The Agrarian Roots of Authoritarian Modernization in Brazil." Ph.D. diss., Massachusetts Institute of Technology, Department of Political Science.
Reis, José. 1976a. "Artur Neiva, o homem e a obra." *Ciência e Cultura* 28, no. 6: 707–12.
———. 1976b. *Grandeza científica de São Paulo.* São Paulo: ACIESP. Publicação ACIESP, no. 1.
———. 1976c. "Instituto Biológico de São Paulo." *Ciência e Cultura* 28, no. 5: 576–601.
———. 1976d. "Rocha Lima, o homem e a obra." *Ciência e Cultura* 28, no. 4: 463–79.
Rheinboldt, H. 1955. "A química no Brasil." In Azevedo 1955, 2:9–89.
Ribeiro, J. Costa. 1955. "A física no Brasil." In Azevedo 1955, 1:163–202.
Ringer, Fritz, K. 1969. *The Decline of the German Mandarins: The German Academic Community, 1890–1933.* Cambridge: Harvard University Press.
Romani, Jacqueline Pitangui. 1982. "O Conselho Nacional de Pesquisas e a institucionalização de pesquisa científica no Brasil." In Schwartzman 1982:137–68.
Rosa, J. N. Santa. 1974. "A formação de um mestre de pesquisa tecnológica (o Núcleo da Estação Experimental de Combustíveis e Minérios)." *Revista de Química Industrial* 501 (January): 2–6.
Rosemberg, H. 1966. *Bureaucracy, Aristocracy, and Autocracy: The Prussian Experience, 1660–1815.* Cambridge: Harvard University Press.
Rothblatt, S. 1985. "The Notion of an Open Scientific Community in Historical Perspective." In Gibbons and Wittrock 1985:21–76.

Rushing, Francis W., and Carole Ganz Brown, eds. 1986. *National Policies for Developing High Technology Industries.* Boulder, Colo.: Westview Press.
Sábato, Jorge, ed. 1975. *El pensamiento latinoamericano en la problemática ciencia. Tecnología, desarrollo, dependencia.* Buenos Aires: Editorial Paidós.
Sagasti, Francisco. 1983. *La política científica y tecnológica en América Latina. Un estudio del enfoque de sistemas.* Mexico: El Colegio de Mexico.
Salem, Tânia. 1982. "Do Centro D. Vital à Universidade Católica." in Schwartzman 1982:97–136.
Sales, Dagoberto. 1958. *Energia atômica. Um inquérito que abalou o Brasil.* Rio de Janeiro: Editora Fulgor.
Salomon, J.-J. 1970. *Science et politique.* Paris: Editions du Seuil.
Santos Filho, Lycurgo. 1947. *História de medicina no Brasil (do século XVI ao século XIX).* 2 vols. São Paulo: Editora Brasiliense.
———. 1977. *História geral da Medicina Brasileira.* São Paulo: Editora Hucitec and Editora da Universidade de São Paulo.
Saraiva, A. José. 1955. *História da cultura em Portugal.* 2 vols. Lisbon: Editora Jornal do Foro.
Schwartzman, Simon. 1973. "Regional Contrasts within a Continental-Scale Nation: Brazil." In Eisenstadt and Rokkan 1973. 2: 209–32.
———. 1975. *São Paulo e o estado nacional.* São Paulo: Difel.
———. 1978. "Struggling to Be Born: The Scientific Community in Brazil." *Minerva* 16 (Winter): 545–80.
———. 1979. *Formação da comunidade científica no Brasil.* São Paulo and Rio de Janeiro: Cia. Editora Nacional and FINEP.
———. 1980. "The Miracle and Its Costs" (review essay). *Latin American Research Review* 15, no. 2:269–73.
———. 1982. *Bases do autoritarismo brasileiro.* Rio de Janeiro and Brasília: Editora Campus and Editora da Universidade de Brasília.
———. 1983. "La burocratización de la tecnología. El caso del Instituto Nacional de Tecnología." In Mora y Araujo 1983:81–134.
———. 1984a. "A árvore da ciência." *Ciência Hoje* 2 (November–December): 70–84.
———. 1984b. "The Focus on Scientific Activity." In Clark 1984:199–232.
———. 1985. "The Quest for University Research: Policies and Research Organization in Latin America." In Wittrock and Elzinga 1985:101–16.
———. 1986a. "Coming Full Circle: A Reappraisal of University Research in Latin America." *Minerva* 24 (Winter): 456–75.
———. 1986b. "A política da igreja e a educação. O sentido de um pacto." *Religião e Sociedade* (Rio de Janeiro), 13 (March): 108–127.
———. 1988a. "Brazil: Opportunity and Crisis in Higher Education." *Higher Education* 17, no. 1: 99–119.
———. 1988b. "High Technology and Self-Reliance: Brazil Enters the Computer Age." In Chacel, Falk & Fleischer 1988:67–82.
———. 1991. "Changing Roles of New Knowledge." In Wagner et al., eds., 1991: 230–60.
Schwartzman, Simon, ed. 1982. *Universidades e instituições científicas no Rio de Janeiro.* Brasília: Conselho Nacional de Desenvolvimento Científico e Tecnológico.
———. 1983. *Estado Novo. Um auto-retrato.* Brasília: Editora da Universidade de Brasília.
Schwartzman, Simon, Helena M. Bousquet Bomeny, and Vanda M. Ribeiro

Costa. 1984. *Tempos de Capanema*. Rio de Janeiro and São Paulo: Paz e Terra and Editora da Universidade de São Paulo.
Schwartzman, Simon, and Cláudio de Moura Castro, eds. 1986. *Pesquisa universitária em questão*. São Paulo: Unicamp/Icone/CNPq.
Schwartzman, Simon, and Maria Helena M. Castro. 1984. "Nacionalismo, iniciativa privada e desenvolvimento industrial. Os primórdios de um debate." *Dados—Revista de Ciências Sociais* 27, no. 1: 89–111.
Secretaria Especial de Informática. 1984. "Panorama da indústria nacional. Computadores e periféricos." *Boletim Informativo* 4 (September).
Sérgio, Antônio. 1972. *Breve interpretação da história de Portugal*. Lisbon: Livraria Sá Costa.
Shaplen, Robert. 1964. *Towards the Well-being of Mankind (Fifty Years of the Rockefeller Foundation)*. New York: Doubleday.
Shils, Edward. 1968. *Criteria for Scientific Development*. Cambridge: MIT Press.
Silva, Maria Beatriz Nizza. 1988. "O pensamento científico no Brasil na segunda metade do século XVIII." *Ciência e Cultura* 40, no. 9: 859–68.
Silva, Maurício Rocha e. 1960. "Dez anos pelo progresso da ciência." *Revista Brasileira de Estudos Pedagógicos* 33 (January–March): 221–34.
———. 1978. "Fundação e história da SBPC. Trinta anos em defesa da ciência." *Ciência e Cultura* 30, no. 10:1183–88.
Simonsen, Roberto. 1962. *História econômica do Brazil*. São Paulo: Cia. Editora Nacional.
Skidmore, Thomas E. 1967. *Politics in Brazil, 1932–1964: An Experiment in Democracy*. New York: Oxford University Press.
Smith, T. Lynn, and A. Marchant. 1951. *Brazil: Portrait of Half a Continent*. New York: Dreyden Press.
Sociedade Brasileira de Física. 1987. *A física no Brasil*. São Paulo: Sociedade Brasileira de Física.
Soper, Fred, and Bruce Wilson. 1943. *Anopheles Gambiae in Brazil, 1930 to 1940*. New York: Rockefeller Foundation.
Souza, A. Cândido de Melo. 1960. "Letras e idéias no Brasil colonial." In Holanda 1960b, 1:91–105.
Spiegel-Rösing, Ina, and Derek de Solla Price, eds. 1977. *Science, Technology, and Society: A Cross-Disciplinary Perspective*. Beverly Hills, Calif.: Sage Publications.
Stein, Stanley. 1957. *The Cotton Textile Industry in Brazil, 1850–1950*. Cambridge: Harvard University Press.
Stepan, Nancy. 1976. *Beginnings of Brazilian Science: Oswaldo Cruz, Medical Research, and Policy, 1890–1920*. New York: Science History Publications.
———. 1984. "Eugenics, Genetics, and Public Health: A Brazilian Connection, 1900–1930," paper presented to the American Historical Association/History of Science Society meetings.
Stols, Eddy. 1974. "Les Étudiants brésiliens en Belgique (1817–1914)." *Revista de História* (São Paulo), 25, no. 100, vol. 2:653–92.
Tigre, Paulo Bastos. 1983. *Technology and Competition in the Brazilian Computer Industry*. New York: St. Martin's Press.
Tobias, J. Antônio. 1968. *História da educação brasileira*. São Paulo: Editora Juriscredi.
Todaro, M. Patrice. 1971. "Pastors, Prophets, and Politicians: A Study of the Brazilian Catholic Church, 1916–1945." Ph.D. diss., Columbia University.

Vale, J. Ribeiro. 1975. "Esboço histórico sobre a farmacologia no Brasil." *O Estado de São Paulo*. Suplemento do centenário, no. 8.
———. 1977. *A Escola Paulista de Medicina*. São Paulo: Revista dos Tribunais.
Velho, Octávio G. 1976. "Modos de desenvolvimento capitalista, campesinato e fronteira em movimento." *Dados* (Rio de Janeiro), 13:15–32.
Velloso, João Paulo dos Reis. 1986. *O último trem para Paris*. Rio de Janeiro: Nova Fronteira.
Venâncio Filho, A. 1977. *Das arcadas ao bacharelismo. 150 anos de ensino jurídico no Brasil*. São Paulo: Editora Perspectiva.
Verney, L. Antônio. 1949–50. *Verdadeiro método de estudar*. 5 vols. Lisbon: Livraria Sá Costa.
Vessuri, Hebe. 1986. "The Universities, Scientific Research, and the National Interest in Latin America." *Minerva* 24, no. 1: 1–38.
———. 1987. "The Social Studies of Science in Latin America." Mimeograph.
Wade, Nicholas. 1985. "Third World: Science and Technology Contribute Feebly to Development." *Science* 189, no. 4205: 770–76.
Wagner, Peter, Carol Weiss, Björn Wittrock, and Hellmutt Wollman, eds. 1991. *Social Science and Modern States*. Cambridge: Cambridge University Press.
Weber, Max. 1958. "Science as a Vocation." In Gerth and Mills 1958:129–58.
Whitaker, A. P., ed. 1961. *Latin America and the Enlightenment*. 2nd ed. (1st ed. 1942). Ithaca, N.Y.: Cornell University Press.
Wirth, John D. 1970. *The Politics of Brazilian Development, 1930–1954*. Stanford, Calif.: Stanford University Press.
Wittrock, Björn, and Aant Elzinga, eds. 1985. *The University Research System: The Public Policy of the Home of Scientists*. Stockholm: Almqvist & Wiksell International.

INDEX

Ab-Sáber, Aziz, 249
Abraham, Henri, 109
Abreu, Alvaro de Paiva, 92n
Abreu, Silvio Fróes, 92n, 93–94, 95, 95n, 255
Academia Brasileira de Ciências, 81, 108, 109, 110, 111, 117, 120
Académie de Sciences de France, 24–25
academies of science: Brazil, 81, 108, 109, 110, 111, 117, 120; France, 24–25; Italy, 155; Portugal, 40, 44
Adler, Emanuel, 216n, 233n, 234n, 235n, 255
Adorno, Sérgio, 57n, 255
Afonso, Pedro, 86
Agassiz, Louis, 91n
agricultural research, 16, 71, 76, 140–45, 154, 154n, 187, 208, 224, 232, 241

agriculture and animal husbandry schools: in Niterói, Escola Superior de Agricultura e Veterinária, 150; in Piracicaba, Escola Superior de Agricultura Luiz de Queiroz, 76, 133, 141–43, 184, 187; in Rio de Janeiro, Escola de Agricultura e Veterinária, 76, 176
Airosa, Plínio, 134
Albagli, S., 205n, 255
Albanese, Giacomo, 135, 185
Albernaz, Paulo M., 128n, 255
Albertin, P., 18, 255
Albertini, Eugène, 121
Alberto, Álvaro. *See* Silva, Álvaro Alberto da Mota e
Alberto, Amandina Álvaro, 110n
Albuquerque, Cavalcante, 156

alcohol, 94, 153, 239
Alden, Dauril, 51n, 255
Alencar, Otto de, 80, 169
Alliance for Progress, 193
Almeida, A., Jr., 58n, 60n, 106, 113n, 131n, 256
Almeida, Alvaro Osório de, 113, 146, 175, 188, 189
Almeida, J. Oliveira, 209n
Almeida, João Rodrigues Pereira de, 53n
Almeida, Miguel Osório de, 175, 188–89
Alves, Francisco Rodrigues, 87, 141n
Alves, Márcio Moreira, 107n, 256
Alvim, Cesário, 86
Alvim, Francisco Cordeiro da Silva e, 57n
Amado, Gilberto, 112n
Amaral, Afrânio do, 84n, 99–100, 100n, 169, 171, 249, 256
Amaral, Antônio José do, 57
Amaral, Irnack Carvalho do, 92n
Americano, Jorge, 203
Andoyer, H., 81n
Andrada, José Bonifácio de. *See* Silva, José Bonifácio de Andrada e
Andrade, José Ferreira de, Jr., 92n
Andrade, Nuno de, 85–86, 86n, 87
animal husbandry. *See* agriculture and animal husbandry schools
Anjos, Ciro dos, 210
Antivenom Institute of America, 99
Apple computers, 234
Aquinas, Thomas, 35, 38
Aragão, Henrique Beaurepaire, 89, 90, 111n
Arens, Karl, 120, 135
Argentina, 99
Argon Laboratory, 158
Aristotle, 3, 24, 34, 35, 38
Associação Brasileira de Educação, 108, 109, 110, 111, 117, 118, 120
Associação Comercial de São Paulo, 127
astronomy, 59, 78–79, 157, 169
Ataíde, Tristão de. *See* Lima, Alceu Amoroso
Athanasov, Nicolas, 141n
atomic energy. *See* nuclear research
Australia, 7
Austria, 97, 140, 142, 173
Azevedo, Fernando de, 10–11, 55, 77n, 106–7, 127n, 131n, 134, 135, 195, 256

Azevedo, Inácio de, 111n
Azevedo, Moreira de, 51, 256
Azevedo, Roberto Marinho de, 81n, 120, 121

Bacha, Edmar, xi, 256
Backheuser, Everardo, 110n
bacteriological research, 72, 76, 83–91. *See also* Instituto Manguinhos; tropical medicine
Bahia, Antonio Valadares, 103n
Balán, Jorge, 48n, 256
Baldus, Herbert, 210n
Baltensberger, Walter, 206
Banco Nacional de Desenvolvimento Econômico (BNDE), 5, 215, 216, 218, 225, 228, 229, 233, 257
Barata, Mário, 59n, 256
Barbosa, Otávio, 103, 169
Barbosa, Plácido, 88n, 256
Barnes, B., 13n, 24, 256
Barreto, Mauro Pereira, 209n
Barros, Ademar de, 148, 161n, 203
Barros, Fernando de Souza, 179, 226
Barros, R. S. Maciel de, 41n, 58n, 256
Barros, Rosina de, 143
Basalla, George, 7n, 256
Bastide, Paul Arbusse, 134
Bastide, Roger, 108n, 135, 256
Batista, Aníbal Teotônio, 103n
Bauer, Peter, 246n, 256
Bayer, Adolph von, 103n
Beck, Guido, 173, 174, 249
Belgium, 77
Bell Telephone Laboratories, 182, 183, 225, 226
Bella, Robert N., 7n, 256
Bellevue Hospital (New York), 192
Ben-David, Joseph, 7, 9n, 11n, 25n, 29n, 30n, 31n, 153–54, 256
Benchimol, Jaime L., 256
Bendix, Reinhard, 6, 257
Beraldo, Wilson, 168n, 178
Berlink, Ciro, 128n, 257
Bernal, J. D., 8n, 25n, 202, 257
Bertels, D. E., 18n, 258
Berulle, Cardinal of, 37n
Berveille, Michel, 134
Beviláqua, Clóvis, 58n
Beviláqua, Luís, 254

Index 271

Bhaba, H., 157
Biblioteca Nacional, 51
Bier, Otto, 99, 145, 146, 171, 175, 250
Bignon, Jean-Paul, 25
biochemistry, 102–3, 147, 178, 209, 211n
biological sciences, 8, 28, 54–55, 71, 109, 112, 130, 133, 134, 137, 139, 142, 144, 145n, 169, 171, 174, 176–77, 179, 181, 185, 187–88, 224
biophysics, 146–48, 187
biostatistics, 209
biotechnology, 232, 241
Birmingham Lunar Society, 26
Bittencourt, Agesilau, 131n
Bittencourt, Paulo, 161
Blackett, Patrick M., 157, 161
Bloor, David, 13n, 257
Blount, J. A., 84n, 257
Boff, Luis Valente, 206, 226
Bohr, Niels, 154
Bolivia, 158
Bomeny, Helena Maria Bousquet, 73n, 117n, 120n, 122n, 123n, 124n, 125n, 126n, 266
Bonzon, Alfred, 135
Borel, Émile, 82, 109
Born, Max, 156n
Borne, Etienne, 134
botanical gardens: in Belém, 51; in Rio de Janeiro (former Real Horto, then Real Jardim Botânico), 51, 52
botany, 27–38, 39–40, 50–54, 65, 132, 134, 144, 178, 187
Botelho, Antonio José, 203n, 257
Bovero, Alfonso, 100
Boxer, C. R., 50n, 257
Brade, Kurt, 168n
Braga, Ernâni, 191n, 192n, 257
Bragg, Sir Henry, 157
Bragg, William Lawrence, 157
Branco, Carlos Humberto Castelo, 212, 217
Branner, John Caspar, 91n
Brasil, Vital, 85, 86, 99, 99n
Brazil. See by institution
Brazilian colonial economy, 47–49
Brazilian economic "miracle," 3
Bréhier, Émile, 121
Brener, Martha, 143
Breslau, Ernest, 134, 135, 187

Brieger, Friedrich Gustav, 142–43, 163, 172–73, 187, 250
Briquet, Raul, 131n
British Association for the Advancement of Science, 26, 31
Brizola, Leonel, 210n
Brown, Carole Ganz, 266
Brumpt, Emilio, 100
Bruneau, Thomas C., 107n, 257
Brunschvig, Leon, 81n
Buffon, George-Louis, 27
Burgher, von, 150
Burtt, Edwin A., 24n, 257
business administration, 228
business schools: in Rio de Janeiro and São Paulo, Escola de Comércio, 76
Butantã Institute. See Instituto Butantã

California Institute of Technology, 183, 206
Calógeras, Pandiá, 92n
Camargo, Aspásia Alcântara de, ix n
Camargo, Teodureto de, 140
Cambridge University, 24, 30, 52, 130, 137, 155, 156n, 157, 180, 183, 186
Campos, Bernardino de, 141n
Campos, Ernesto de Souza, 63n, 83n, 112n, 135n, 257
Campos, Francisco, 114–18, 119, 122, 123, 124, 129–31, 133, 137, 140n
Campos, Luís Felipe Gonzaga, 91n, 92, 93
Campos, Melo, 103n
Capanema, Gustavo, 114n, 122–26, 127, 137, 190, 197n
Carajás mining complex, Brazil, 239
Cardoso, Vicente Licínio, 80, 81, 111n, 112n, 113n
Cardwell, D. S. L., 25n, 257
Carmichael, H., 157
Carneiro, J. F. D., 219n, 257
Carneiro, Levi, 111n
Carneiro, Mário Barbosa, 92–93n
Carnot, Sadi, 28
Carone, Edgard, 87n, 257
Carrero, Júlio Porto, 110n
Cartaxo, Ernâni, 113n, 257
Cartesianism, 68. See also Descartes
Carvalho, Arnaldo Vieira de, 100–101, 191
Carvalho, Carlos Gregório de, 110n
Carvalho, Hervásio G., 233n, 257

Carvalho, J. de, 38n, 258
Carvalho, José Murilo de, x, 42n, 58n, 61n, 62n, 88n, 258
Carvalho, Leôncio de, 58
Carvalho, Paulino Franco de, 92
Casa dos Pássaros, 52
Castro, Almir de, 212, 250
Castro, Antônio de Barros, 50n, 258
Castro, Antônio Oliveira, 146
Castro, Cláudio de Moura, 200, 223n, 240n, 258, 267
Castro, Francisco Mendes de Oliveira, 57n, 59n, 60n, 61n, 82, 83, 120, 162, 163, 217n, 258
Castro, Maria Helena de Magalhães, 92n, 267
Catholic church, 41–43, 72; Catholic movement for education, 106, 107, 108, 117, 119, 122, 126; Congregação do Oratório, 37; Dominicans, 37; Jesuits, 7, 34–38, 41–43, 49, 50, 73
Catholic University. See Universidade Católica
Catunda, Omar, 134
Cauchy, 82
Cava, Ralph della, 107n, 108n, 258
Cavendish Laboratory, 173
Center for Advanced Study, Princeton, 157
Central Europe, 7, 34
Centre National de la Recherche Scientifique (CNRS), 190
Centro Brasileiro de Pesquisas Educacionais, 210n, 219
Centro Brasileiro de Pesquisas Físicas (CBPF), 162, 203–4, 225
Centro de Pesquisa e Documentação em História Contemporânea do Brasil (CPDOC), Rio de Janeiro, ix n, 16n, 123n, 258
Centro de Tecnologia do Estado (former Instituto de Tecnologia Industrial), Minas Gerais, 103n
Centro Dom Vital, Rio de Janeiro, 107
Centro Tecnológico da Aeronáutica (later Centro Técnico Aeroespacial), 205, 208, 216
Centro Tecnológico de Informática, São Paulo, 235, 236
Cerelli, Francesco, 155
Chacel, Julien M., 258

Chagas, Carlos, 86, 89, 98, 102n, 174, 187, 188, 192
Chagas, Carlos Pinheiro, 102, 191
Chagas Filho, Carlos, 146, 147, 148, 168n, 174, 175, 176, 178n, 188, 197, 250
Chagas Filho, Evandro, 147, 174, 188, 192
Chandrasekhar, Subrahmanyan, 157
Chapot-Prévost, E., 86n
Chateaubriand, Assis, 189
chemistry, 8, 16, 28, 39, 45, 51, 53, 59, 64, 65, 109, 112, 125, 132, 133, 134, 137, 142, 144, 148–54, 179n, 180, 182–83, 191, 193, 216, 228
chemistry schools: in Minas Gerais, Instituto de Química at the Escola de Engenharia de Belo Horizonte, 103; in Rio de Janeiro, Escola Nacional de Química (later Instituto de Química), 150, 179n, 180, 182, 227
Chur, L. A., 18n, 53n, 258
Cidade, Hernani, 258
Cirurgião-Mór do Exército, 63
Clark, Burton R., 258
Clebsh, Alfred, 80
Cobra S.A., 234
coffee, 48, 72, 73
Coimbra, Alberto Luís, 227, 228–29, 250
Colbert, Jean-Baptiste, 24, 25
Colégio Arnaldo, Minas Gerais, 102
Colégio das Artes, Portugal, 35–36, 37
Colégio dos Nobres, Portugal, 38
Colégio Pedro II, Rio de Janeiro, 102
Collor, Fernando. See Melo, Fernando Collor de
Colombia, 191
colonial heritage, 7, 18, 23, 32–46, 49–51. See also Portuguese colonization
Columbia University, 106, 119, 177; Teachers College, 106
Comissão de Energia Atômica, Brazil, 204
Comissão de Exploração Geográfica e Geológica de Minas Gerais, 91n
Comissão dos Estudos de Carvão, 91n
Comissão Especial do Planalto Central do Brasil, 91n
Comissão Geográfica e Geológica de São Paulo, 91n, 140
Comissão Geológica do Império. See Departamento Nacional da Produção Mineral

Companhia Docas de Santos, São Paulo, 175n, 188
Companhia Hidrelétrica do São Francisco (CHESF), 92n
Compton, Arthur, 155, 158, 160
computer science and technology, 232–36, 241
Comte, Auguste, 75, 79, 80. *See also* positivism
Congress, Brazil, 87, 234, 257
Conniff, Michael, 119n, 258
Conselho de Segurança Nacional, 234
Conselho Federal de Educação, 209, 213, 223
Conselho Nacional de Desenvolvimento Científico e Tecnológico. *See* Conselho Nacional de Pesquisas
Conselho Nacional de Educação, 116
Conselho Nacional de Pesquisas (CNPq), xi, 189, 204, 205, 216, 218, 239, 241
Constant, Benjamin. *See* Magalhães, Benjamim Constant Botelho de
Constitutionalist Revolution (São Paulo, 1932), 127–29
Conversi, Marcello, 161, 162n
Coordenação de Aperfeiçoamento de Pessoal de Nível Superior (CAPES), 205, 222, 223, 240
Coordenação dos Programas de Pós-Graduação em Engenharia (COPPE), 97, 216, 223, 227–33
Coornaert, Émile, 134
Copernicus, 25
Corbusier, Le, 124
Cordeiro, Antônio, 184, 212
Cornell University, 91n, 141
Correns, Karl E., 142
cosmic rays research, 94, 156, 157, 158
Costa, Borges da, 102
Costa, Cláudio, Neto, 250
Costa, Domingos, 79n
Costa, Ernesto Luís da Fonseca, 93, 94, 189
Costa, Fernando, 148
Costa, Lúcio, 124
Costa, Macedo (Bishop of Belém), 43n
Costa, Manoel Amoroso, 71, 80, 81, 82, 109n, 110n, 111n, 169, 175n, 179n, 258
Costa, N., 192n, 264
Costa, Vanda M. Ribeiro, 73n, 117n, 120n, 122n, 123n, 124n, 125n, 126, 266

Coutinho, Azeredo, 42
Couto, Miguel, 86, 101
Couty, Louis, 55
Covian, Miguel, 209n
Cropani, Ottorino de Fiori, 135
Crosland, Maurice, 25n, 258
Crowell, Bowman C., 192
Cruls, Louis, 78n
Cruz, Oswaldo Filho, 174
Cruz, Walter Oswaldo Filho, 168n, 174, 188n
Cruz, Isar Hasselman Oswaldo, 168n, 258
Cruz, Oswaldo, 85–89, 98, 102, 145n, 169, 174
Cuba, 87
Cunha, Almeida, 102
Cunha, Domingo, 111n
Cunha, Raul Leitão da, 111n
curandeiros (herbal healers), 63, 68
Curie, Marie, 76, 190
Czechoslovakia, 97, 149, 180

Dafert, F. W., 140
Dahlman, Carl J., 232n, 258
Daland, Robert T., 195n, 258
D'Alembert, 26
d'Alessandro, Alexandre, 82n, 258
Dalton, John, 28
Damy, Marcelo. *See* Santos, Marcelo Damy Souza
Danon, Jacques, 76n, 150n, 182, 250
Dantas, Francisco Clementino San Tiago, 126
Dantec, Felix Le, 84
Darboux, Gaston, 80
Darling, S. T., 191
Darwin, Charles, 24, 27, 143
Darwinism, social, 74
Daubrée, Auguste, 61
Dean, Warren, 140n, 258
Debye, P. J. W., 156
Dedijer, Stevan, 259
Deffontaines, Pierre, 121, 126n, 134
Delfim Neto, Antônio, 73, 218, 225, 259
dentistry, 76, 124n, 133
Departamento de Administração do Serviço Público (DASP), 98, 147, 195, 196, 197n
Departamento Nacional da Produção Mineral (formerly Serviço Geológico e Min-

eralógico), 54, 91–92, 95–97, 120, 150n, 170, 196
Departamento Nacional de Saúde Pública, 191
Derby, Orville A., 91–92, 140
Descartes, Renée, 14, 25, 28, 34, 37, 80. *See also* Cartesianism
Deutsche Naturforscher Versammlung, 31
Dias, Cândido da Silva, 156
Dias, Emanuel, 174
Dias, Ezequiel, 86, 89, 102–3, 169, 174
Dias, Mário Ulysses Viana, 176, 192n
Diderot, 26
Digibrás, 234
Dirac, 156n, 157n, 186
Diretoria Geral de Pesquisas Científicas, 94, 189
Diretoria Geral de Saúde Pública, 87
Djerassi, Carl, 154n, 250
Döbereiner, Johana, 168n
Dobzhansky, Theodosius, 143–44, 163
Domingues, Mário, 36n, 37n, 259
Domingues, Otávio, 141
Dreyfus, André, 131n, 133n, 134, 142–43, 178, 179, 187, 208n
Duarte, Paulo, 129–30, 131n, 132n, 133n, 135, 137, 250, 259
Ducke, Adolfo, 168n
Duerk, Hermann, 145n
Dufay, Charles F., 28
Dumas, Georges, 109, 125–26, 135, 136
Dumont, Alberto Santos, 81n, 111n

earth sciences, 8, 28, 40, 45, 51–52, 54, 55, 59, 61, 62, 67, 71, 72, 91–97, 103, 109, 121n, 132, 134, 137, 139, 169, 178
East, Edward M., 142
École de Mines de Saint Etienne, 62
École des Mines de Paris, 61, 137. See also *grandes écoles*
École Nationale de Ponts et Chaussées, 29
École Normale Supérieure, 84n, 137
École Polytechnique, 68, 137, 182, 207
Ecuador, 191
Edinburgh Philosophical Society, 26
educational reforms in Brazil: the Francisco Campos' Reform, 113–14; in 1968, 219–23; the Maximiliano Reform, 113
Einstein, Albert, 81, 109
Eisenstadt, S. N., 259

Electric Research Association, London, 173
Empresa Brasileira de Aeronáutica (EMBRAER), 208
engineering schools: in Minas Gerais, Escola de Engenharia de Itajubá, 103; Escola de Engenharia de Minas Gerais da UFMG, 103, 149, 150, 169; Escola de Minas de Ouro Preto, x, 61–62, 68, 91, 92, 93, 95, 96, 102, 103, 170, 179n; in Pernambuco, Escola de Engenharia de Recife, 150; in Rio de Janeiro, Faculdade de Engenharia (previously Academia Militar, Escola Central, Escola Politécnica), 51, 55, 59–62, 67–68, 78–81, 82, 92, 93, 94, 101, 109, 110, 111, 120, 121, 146, 148, 162, 163, 169, 170, 179; Instituto Militar de Engenharia, Rio de Janeiro 207, 216; in Rio Grande do Sul, Escola de Engenharia de Porto Alegre, 76, 149, 150; in São Paulo, Escola de Engenharia Industrial, 227; Escola de Engenharia Mackenzie, 76; Escola Politécnica da USP, 67, 68, 76, 81–83, 127, 133n, 134, 137, 150, 152, 156, 158, 180, 182, 185, 186, 233
England. *See* United Kingdom
Erber, Fábio, 246n, 259
Ernesto, Pedro, 119
Escola. *See by field:* medicine, engineering, agriculture, etc.
Escola de Aplicação, 60
Escola de Sociologia e Política de São Paulo, 128, 193, 210n
Escola Livre de Sociologia. *See* Escola de Sociologia e Política de São Paulo
Escola Politécnica. *See* engineering schools
Estação Agronômica de Campinas, 148
Estação Experimental de Combustíveis e Minérios, Rio de Janeiro, 93. *See also* Instituto Nacional de Tecnologia
Estado Novo, 72, 114n, 118, 136
estrangeirados (foreignized), 38
ethnography, 53n, 135, 178
ethnology, 132, 178
Evans, Peter B., 235n, 259
Experimental Chemical-Agricultural Station, Vienna, 140

Faculdade. *See by field:* medicine, engineering, agriculture, etc.

Faculdade de Direito, São Paulo, 133n, 134, 137
Faculdade de Filosofia. *See* science and education schools
Faculdade de Higiene e Saúde Pública de São Paulo (previously Instituto de Higiene), 191
Faculté de Lettres, Paris, 81n
Fajardo, Francisco, 86n
Falcão, Edgard Cerqueira, 44n, 45n, 66n, 259
Falcon, Francisco José Calazans, 38n, 259
Falk, Pamela S., 258
Fantappié, Luigi, 125, 134, 155, 156, 160, 185, 186, 206
Faoro, Raymundo, 40n, 259
Faraday, Michael, 28
Faria, José Gomes de, 89, 90
Faria, L. de Castro, 55, 259
Faria, T., 18, 255
Faria, Vilmar, 200, 259
Farquhar, Percival, 93
Fausto, Boris, 259
Federação das Indústrias de São Paulo (FIESP), 127, 152
Feigl, Fritz, 96, 149, 150n
Felipe, Carneiro, 146, 150n
Fermat, Pierre de, 24
Fermi, Enrico, 155, 157, 186
Ferranti computers, England, 234
Ferrari, A. Trujillo, 128n
Ferreira, Alexandre Rodrigues, 50
Ferreira, Cícero, 102n
Ferreira, Clemente, 86n
Ferreira, José Pelúcio, x, 216, 218, 225, 254
Ferreira, Paulo Leal, 182, 250
Ferreira, Ricardo, 179, 181, 183, 250
Ferreira, Sebastião Virgílio, 103n
Ferri, Mário Guimarães, 18n, 168n, 259
Fialho, Branca Osório de Almeida, 110n
Figueiredo, Carlos Burle, 192
Figueiredo, João Batista, 218, 236
Financiadora de Estudos e Projetos (FINEP), ix–x, 5, 216, 218, 225, 241
Finlay, Carlos Juan, 87
Físico-Mór do Reino, 63
Flammarion, Camille, 79
Fleisher, David V., 258
Folin, Otto, 102n

Fonseca, Fernando Valadares, 232n
Fonseca Filho, O., 90n, 171, 175n, 192n, 259
Fonseca, Pedro da, 34
Fontes, Cardoso, 86, 89, 98
Ford Foundation, xi, 211
Fourier, J. B. J., 28
Franca, Leonel, 35n, 259
France, 24–25, 27, 29, 30, 34, 43, 49, 61, 62, 68, 80, 81n, 84n, 86, 87, 109, 137, 151, 176, 182, 190, 207, 226n, 228
Francisco Campos' Reform, the, 113–14. *See also* educational reforms
Franken, Tjerk, x
Franklin, Benjamin, 28
Freemasons, 42, 43n
Freire, Luís de Barros, 121, 156, 179, 181
Freitas, Francisco J. de, 91n
Freitas, J. L. Pedreira de, 209n
French Revolution, 27
Frieiro, Eduardo, 101n, 259
Frischtak, Cláudio, 235n, 259
Fromont, Pierre, 134
Fulbright Commission, 228
Funaro, Dilson, 225
Fundação de Amparo à Pesquisa do Estado de São Paulo (FAPESP), 203, 226n
Fundação Gaffrée-Guinle, 83n, 188
Fundação Getúlio Vargas, ix, 16, 123n
Fundação Instituto Brasileiro de Geografia e Estatística (FIBGE), 3n, 259
Fundo Nacional de Tecnologia (FUNTEC) (replaced by Fundo Nacional de Desenvolvimento Científico e Tecnológico [FNDCT]), 215, 216, 228, 229
Furtado, Celso, 18n, 259
Furtado, Jacundino, 113n, 259

Gaffrée, Cândido, 83n, 175n, 188
Gagé, Jean, 135
Galileo, 23–24, 25
Gall, Norman, 233n, 260
Galvani, Luigi, 28, 135
Gama, Lélio, 79, 81, 120, 169, 170, 251, 260
Gama, Reinaldo Saldanha da, 135
Gama, Vasco da, 33
Gamov, George, 157
Gardner, George, 161

Garfield, E., 240n, 260
Garric, Robert, 121, 126, 134
Gassendi, Pierre, 24
Gauss, Carl, 75
Geiger, J. H. W., 156n
Geiger-Müller counters, 157n
Geisel, Ernesto, 4, 201
genetics, 141–45, 184, 197, 193, 208n
Geological Society of London, 44
Geological Survey, U.S., 97n, 169
geology, 51, 54, 55, 59, 61, 91–97, 169. *See also* earth sciences
George Washington University, 157
Gergsakademie, Freiburg, 29
Germany, 29, 31–32, 34, 44, 49, 87, 90, 94, 97, 100, 103n, 140n, 142, 145, 148, 149, 151–54, 162, 173, 204, 228, 233
Gerth, H. H., 260
Gibbons, M., 13n, 260
Giddens, Anthony, 8n, 260
Giemsa, Gustav, 90
Giesbrecht, Ernesto, 168n, 182, 251
Gilpin, Robert, 25n, 30n, 260
Ginásio Mineiro, 102
Ginásio Petrópolis, 82n
Giorgi, Giovanni, 125
Glaser, William A., 4n, 260
Gley, Emil, 109
Godinho, V. M., 50n, 260
Godói, Alcides, 89, 90
Godoi, Manuel Pimentel, 103n
Goes, Paulo de, 188n
Goethe, 28
Goldemberg, José, 161n, 183, 251
Goldhaber, Maurice, 161
Göldi, Émil, 53, 54
Gomes, Diogo, 33n
Gomes, Francisco de Paula Magalhães, 102, 169, 170, 251
Gomes, João Florêncio, 99n
Gomide, Elza F., 9n, 160
Gonçalves, Cylon E. Tricot, x n, 254
Gonçalves, Francisco Rebelo, 134
Gonçalves, José Moura, 147, 168n, 178, 209n
Gonçalves, Veiga Sales de Moura, 197n
Gorceix, Claude Henri, 61
Gothsch, Otto, 134
Gottlieb, Otto, 16, 154n, 168n, 180, 212, 251

Goulart, João, 210n, 212
Gould, S. J., 28n, 260
Graaf, Van der, 161
Graciarena, Jorge, 202n, 260
graduate education, 5, 32, 216, 218, 221–24, 242
Graham, Douglas, 48n, 260
Graham, Richard, 74n, 260
grandes écoles, 30, 137. *See also under* École
Gray, Stephen, 28
Great Britain. *See* United Kingdom
Grosklauss, Diane I., xi
Gross, Bernard, 94, 120, 147, 162–63, 173, 174, 179, 206, 251
Guerra, E. Sales, 86n, 260
Guerreiro, Cesár, 192
Guilherme, Olímpio, 233n, 260
Guimarães, Cláudia, 254
Guimarães, Djalma, 96, 103n, 120
Guimarães, Manuel Ferreira de Araújo, 57n
Guimarães, Mário Alves, 206
Guinle, Eduardo, 83n, 188
Guinle group, 95, 147, 175
Guinle, Guilherme, 188
Gusmão, Alexandre de, 37

Hamburger, Amélia Império, 181n, 263
Hartmann, Johannes Franz, 90
Hartt, Charles F., 54, 91n
Harvard University, 100n, 102n, 142, 171, 184
Hashimoto, U., 66n, 260
Hauptmann, Heinrich, 135, 149, 151, 153n, 154
Hauser, Henri, 121
Heisenberg, Werner, 156n
Heitler, Walter, 156n, 157
Henrique, Prince, 33
Hermite, Charles, 80
Herrera, Amílcar, 7n, 260
high education reform of 1968, in Brazil, 219–23. *See also* educational reforms
Hime, Eugênio, 162
Hobbes, Thomas, 25
Hoff, J. H. Van't, 149n
Höhne, Frederico Carlos, 168
Holanda, Sérgio Buarque, 18n, 260
homeopathic medicine, 55, 69
Höning, Chaim S., 9n, 260

Hourcade, Pierre, 134
humanities. *See* social sciences and humanities
Hussak, Eugène, 91, 92
Hutton, James, 27

Ignatius of Loyola Society of Jesus. *See* Catholic church
Ihering, Hermann von, 53, 54
Ihering, Rudolfo von, 208n
Imperial University of Tokyo, 66
India, 67, 157, 240, 246
Indian populations, 7, 40, 48, 73, 135, 136
industrial revolution, 25, 26
industrial technology, 241
Inquisition (or Santo Ofício), 7, 35, 36. *See also* Catholic church
Institute of Chemistry, Bonn, 151n
Institute of Tropical Medicine, Hamburg, 145
Instituto Acadêmico do Rio de Janeiro, 45
Instituto Agronômico de Campinas, 76, 131n, 141–42, 143, 260
Instituto Agronômico do Norte, Belém, 154
Instituto Bacteriológico (later incorporated as Instituto Butantã; later renamed as Instituto Adolfo Lutz), 76, 84, 85
Instituto Bacteriológico de Buenos Aires, 99
Instituto Biológico de São Paulo (formerly Instituto Biológico de Defesa Agrícola e Animal de São Paulo), 98, 99, 127, 131n, 144–46, 147, 148, 171, 174, 176, 187, 208
Instituto Butantã, São Paulo, 76, 85, 98, 99, 100, 127, 148n, 169
Instituto de Biofísica, Universidade do Rio de Janeiro, 146, 147, 148, 187, 188, 197, 260
Instituto de Eletrotécnica, São Paulo, 158n, 160
Instituto de Estudos Sociais e Políticos de São Paulo (IDESP), 9n
Instituto de Matemática Pura e Aplicada, Rio de Janeiro, 228
Instituto de Organização Racional do Trabalho (IDORT), 127–28
Instituto de Pesquisas e Desenvolvimento, São José dos Campos, 206

Instituto de Pesquisas Radioativas, Minas Gerais, 204
Instituto de Pesquisas Tecnológicas (IPT), São Paulo, 82n, 160, 208
Instituto de Química Agrícola, Rio de Janeiro, 154, 168n
Instituto de Tecnologia Elétrica, Faculdade de Engenharia do Rio de Janeiro, 146
Instituto de Tecnologia Industrial, Minas Gerais, 103n
Instituto Ezequiel Dias, Minas Gerais, 102–3
Instituto Franco-Brasileiro de Alta Cultura, 109
Instituto Henrique Kopke, Rio de Janeiro, 81n
Instituto Manguinhos (same as Instituto Oswaldo Cruz), 97–103, 120, 136, 144–48, 164, 169, 171, 172, 174–76, 178, 179, 187, 188–89, 192, 197, 238
Instituto Nacional de Tecnologia, 94, 120, 121, 147, 162, 163, 174, 189
Instituto Soroterápico Federal (later Instituto de Patologia Experimental, Manguinhos), 88
Instituto Soroterápico Municipal, Rio de Janeiro, 86, 87
Instituto Superior de Estudos Brasileiros (ISEB), 216
Instituto Vacinogênico (later incorporated as Instituto Butantã), 76, 83, 84, 101
Instituto Vital Brasil, 99. *See also* Brasil, Vital
Inter-American Development Bank, 243
International Business Machines (IBM), 234, 236
International Hygiene Exposition of 1907, Berlin, 88
International Monetary Fund, 240
Itaipu hydroelectric power plant, 239
Italy, 82n, 117, 124, 155, 162, 172, 173

Jacob, Gerhard, 183, 251
Janet, Paul, 109
Japan, 7, 66–67
Japan, universities in, 66
Japanese migration, 7, 19
Jardim Botânico. *See* botanical gardens
Jesuits. *See* Catholic church
João I, King, 33

João V, King, 37, 38
João VI, King, 45, 51n, 52, 59
John Innes Institute, England, 142
Johns Hopkins School of Hygiene and Public Health, 191
Johns Hopkins University, 182, 191, 225
Joliot-Curie, Frédéric, 202
Jordan, Camille, 80
Jordan, Ernest P., 156
Jordão, Pacheco, 91n
Jornal do Brasil, 261
José I, King, 38–39

Kaiser-Wilhelm-Gesellschaft, 32, 142
Kapitza, Peter L., 156
Katz, Jorge, 261
Kepler, J., 25
Kerr, Warwick, 184, 251
Kerst, Donald W., 161
Khulman, João Geraldo, 168n
Klauss, Rudolph, 99
Klein, Herbert S., xi
Klobusitzki, Dionisius von, 100n
Klotz, Oscar, 191
Knorr-Cetina, Karin, 5n, 13n, 14, 261
Köberle, Fritz, 209
Koizumi, K., 66n, 261
Komissarov, B., 18n, 258
Köning, Paul, 100n
Könings, Gabriel, 80
Krombholz, Pawel, 149
Krug, Carlos A., 141, 142n, 143
Kubitschek, Juscelino, 210
Kuhn, Thomas S., 6, 8n, 12n, 261

Laboratório da Rua Bahia, Minas Gerais, 103n
Laboratório de Análise do Estado, Belo Horizonte, 148–49
Laboratório de Biofísica (later Instituto de Biofísica), Rio de Janeiro, 187
Laboratório Nacional da Produção Mineral, Rio de Janeiro, 149, 150n, 179n, 189
Laboratório Paulista de Biologia, 174
Laboratório Químico-Prático, 51n
Laboriau Filho, Ferdinand, 92n, 110n, 111n, 112n, 113n, 261
Lacaille, O., 60n
Lacaz, Carlos da Silva, 63n, 261

Lacerda, Cândido de, 38n
Lacerda, João Batista de, 53n, 54n, 55, 86n, 261
Lacombe, Américo Jacobina, 41n, 261
Lacombe, Laura Jacobina, 110n
Lacroix, Alfred F., 60n
Ladosky, Waldemar, 168n, 178n, 261
Lagrange, Joseph-Louis, 60n
Lambert, Jacques, 126n
Lambert, Robert, 191n
Lang, James, 18n, 50n, 261
Langsdorff, G. I., 18n, 53
Laplace, Pierre S., 60n
Larrabure, Fernando Jorge, 135
Latour, Bruno, 5n, 13n, 14n, 33n, 261
Lattes, Cesare, 161–62, 182, 186, 204, 225, 251
Lavoisier, A. L., 26, 28
Lavradio, Marquis, 51
law, 17, 24, 29, 57; jurisprudence, 45; law schools, 76, 102, 103–4, 117, 131n, 134n, 137
Leão, Aristides Pacheco, 168n
Lebesgue, Henri-Léon, 82
Léduc, Gaston, 121
Leff, Nathanael, 160n, 261
Leibniz, G. W., 26, 28, 34
Leinz, Viktor, 91n, 96, 120, 121n, 122n, 173, 196, 251, 261
Leite, Rogério C. Cerqueira, 183, 206, 225, 226, 233n, 251, 261
Leme, Alberto Betim Pais, 92n, 109n
Leme, Cardinal, 107
Lent, Herman, 98n, 120, 177, 251
Leonardos, Othon, 51, 91n, 92n, 110, 111n, 169, 170, 261
Lessa, Carlos, 217n, 261
Lévi-Strauss, Claude, 134, 135, 136, 237, 238, 261
Levy, Daniel C., 214n, 261
Liais, Emmanuel, 78
Licenko, N., 18n, 258
Liceu de Artes e Ofícios, São Paulo, 160
Liceu de Ouro Preto, Minas Gerais, 102
Lima, Alceu Amoroso (Tristão de Ataíde), 107, 114n, 122, 125n, 126
Lima, Angelo da Costa, 89n, 168n
Lima, Elon Lages de, 212, 254
Lima, Henrique Rocha, 89, 99, 131, 133n, 144, 145, 146, 147, 148

Lima, Silva, 65–66
Linnaeus, C. von, 27
Lins, Ivan M. de Barros, 75n, 262
Lira, Heitor, 106
Lisboa, Marques, 86, 102
Lisboa, Miguel Arrojado, 92n
Lison, Lucien, 209n
Lister, J., 84n
Littré, Émile, 75
Lobachevsky, Nicolay, 75
Lobato, Monteiro, 95, 96, 262
Lobo, Francisco Bruno, 56, 63, 64, 65, 113, 114, 116, 117, 262
Lodi, Jurandir, 209
Löfgren, Alfredo, 109n
Lopes, Hildefonso Simões, 93
Lopes, Hugo de Souza, 177, 251
Lopes, José Leite, 162n, 179, 180, 181, 204, 216, 217, 252, 262
Lopes, Luís Simões, 147, 195, 197n
Lopes, Tito Enéias Leite, 147
Louisiana State University, 180
Loureiro, Ângela, 68n, 69n, 83n, 262
Lourenço Filho, M. B., 106
Louvois, Camille Letellier, 25
Lutz, Adolfo, 84–85, 86, 99
Luz, Rogério, 68n, 69n, 83n, 262
Lyell, Charles, 27

Machado, Freitas, 150n
Machado, João, 147
Machado, Roberto, 68n, 69n, 83n, 262
Magalhães, Benjamin Constant Botelho de, 75, 92n, 149n
Magalhães, C., 37n, 262
Magalhães, Fernando, 63n, 262
Magalhães, Otávio, 103n
Maia, Paulo Castro, 111n
Majorana, Ettore, 156n
Malamphy, Mark C., 96
Malebranche, N., 37n
Manchester, Alan K., 74n, 262
Manchester Literary and Philosophical Society, 26
Manchester University, 182
Manhattan Project, 13
Manifesto dos Pioneiros da Educação Nova (Manifest of the Pioneers of New Education), 106
Maniolescu, Mihail, 117

Marchant, Alexander, 51n, 262
Marchant, Anyda, 51n, 262
Marchouy, Émile, 109
Marcus, Ernst, 135, 168n, 187
Marcus, Evenine, 168n
Mariani, Maria Clara, x, 147n, 219n, 262
Maritain, Jacques, 126n
Martin V, Pope, 33
Martins, Amilcar Filho, 49, 262
Martins, Amílcar Viana, 103n, 171, 252
Martins, Ermínio, xi
Martins, Roberto B., 49n, 262
Martins, Tales, 84n, 100, 175n, 262
Marxism, 75
Mascarenhas, Sérgio, 163, 226
Mason, S. F., 24n, 25n, 26n, 31n, 262
Massachusetts Institute of Technology (M.I.T.), 206
Massarani, Giulio, 227, 228n
mathematics, 9n, 37n, 39, 45, 59–60, 61, 68, 72, 75, 78–81, 83, 109, 112, 125, 132, 133, 134, 135, 156, 169
Mathias, Simão, 148n, 150n, 151, 152–53, 180, 262
Matos, Heraldo de Souza, 94n
Max Planck Institutes, 32
Maximiliano Reform, 113. *See also* educational reforms
Maxwell, James Clerk, 28
Maxwell, Kenneth, 50n, 262
McCann, Frank D., Jr., 262
McLeod, R. M., 7, 50n, 262
medical schools: in Bahia, Escola de Medicina (previously Colégio Médico-Cirúrgico, then Escola de Anatomia e Cirurgia), 51, 63–66, 69, 84n, 145n; Escola Tropicalista Bahiana, 65–66; in Minas Gerais, Faculdade de Medicina de Belo Horizonte, 102–3, 171, 178, 191; in Pará, Faculdade de Medicina, 176; in Pernambuco, Faculdade de Medicina, 176; in Ribeirão Preto, Faculdade de Medicina, 208, 210; in Rio de Janeiro, Escola Médico-Cirúrgica, 51, 52, 64–66, 84n, 86; then, Faculdade de Medicina, 100, 145n, 146, 171, 176, 179, 187; in São Paulo, Escola Paulista de Medicina, 128, 240, 243; Faculdade de Medicina, 98, 100, 127, 128, 134, 137, 152, 160, 172, 176, 191, 192, 194, 196, 208

medical societies, 64, 69, 85n
medicine, x, 8, 17, 24, 28–29, 63–66, 68, 76–77, 131n, 194, 224
Meiller, J. Luís, 82n, 262
Melo, Antônio Manuel de, 78n
Melo, Fernando Collor de, 233
Melo, J. A. G., 50n, 262
Melo, Sebastião José de Carvalho e, 38n. *See* Pombal
Melo, Silva, 175n
Mendel, L. B., 102n
Mendel's law, 142
Mendes, Carlos Teixeira, 141
Mendonça, Siqueira, 110n
Merton, Robert K., 5n, 12, 25n, 262
Mesquita Filho, Júlio de, 127, 129n, 130, 131n, 132n, 133n, 134n, 135, 136, 137, 263
Metallurgical Laboratory, U.S., 158
metallurgy, 40, 59
Mexico, 7n
Meyer, Herta, 147, 197n
Meyer, João Alberto (Jean), 182
Meyer, Víctor, 148
Miceli, Sérgio, 9n, 263
Michler, Wilhelm, 148
midwifery, 65
military government, Brazil, 3–4, 200–201, 239–40
military research and technology, 59, 154, 158–60, 205, 208, 216, 232, 239, 241
military schools, Brazil: Academia Militar do Rio de Janeiro (later Escola Politécnica), 51, 52, 57, 59, 60, 67, 78; Academia Real de Marinha, Rio de Janeiro, 51, 59; Instituto Militar de Engenharia, Rio de Janeiro, 207, 216; Instituto Tecnológico da Aeronáutica, São José dos Campos, 183, 205–7, 212, 225, 226, 233, 243
Miller, Harry M., 143, 193
Millikan, Robert A., 155
Mills, C. H., 260
mineralogy. *See* earth sciences
Mineralogy Society of Jena, 44
Mingóia, Quintino, 173, 174, 252
Mining Code, Brazil, 95
Ministério da Aeronáutica, 206–7
Ministério da Agricultura, Indústria e Comércio, 79, 92, 93, 94, 95n, 140, 141n, 149, 154, 188

Ministério da Educação e Saúde, 114, 117, 122, 124n, 136, 196, 205, 206, 207, 209, 213, 219, 222, 244
Ministério da Fazenda, 218
Ministério da Marinha, 92
Ministério de Ciência e Tecnologia, 240–41, 263
Ministério de Relações Exteriores, 157
Ministério do Trabalho, Indústria e Comércio, 94
Modern Art Week of 1922, São Paulo, 108
Monbeig, Pierre, 134
Monteiro, Hugo Jorge, 252
Monteiro, Lemos, 100
Monteiro, Sérgio Neves, 254
Montenegro Filho, Casimiro (Brigadier), 206, 207, 254
Monteux, Yolande, 156
Mora y Araujo, M., 263
Morais, Abrãao de, 156, 263
Morais, José de Melo, 141n, 142
Morais, Marcílio, x n, 57n, 60n, 78n, 79n
Moravcsik, Michael J., 245n, 263
Morehause, W., 67n, 263
Moreira, Juliano, 109n
Moreira, Manuel da Frota, 168n, 177, 189, 250
Moreira, Oromar, 147
Morel, Carlos, 240n, 263
Morel, Regina L. Morais, 233n, 240n, 263
Morize, Henrique, 78n, 79, 109, 162n, 169, 263
Morner, M., 263
Mors, Walter B., 151n, 154, 168n, 174n, 180, 252, 263
Moscoso, Tobias, 111, 112, 113
Moses, Artur, 109
Mota, Ivone Freire, 181n, 263
Motoyama, Shozo, 259
Mougé, Jean, 134
Moura, Pedro de, 92n
Moussatché, Haiti, 168n
Moyal, A. M., 50n, 263
Mulkay, M. J., 12n, 263
Müller, Fritz, 53
Muricy, Kátia, 68n, 69n, 83n, 262
Murtinho, Joaquim, 55
Museu de Astronomia e Ciências Afins, 79n, 263
Museu do Pará, Brazil, 54

Museu Imperial (former Museu Real, later Museu Nacional), Rio de Janeiro, 51, 52
Museu Nacional, Rio de Janeiro, 52, 53, 54, 148, 172, 175, 263
Museu Paraense Emílio Goeldi, 76, 263
Museu Paulista, 54, 76
Museu Real (later Museu Imperial and then Museu Nacional), Rio de Janeiro, 52
Mussolini, Benito, 172

Nachbin, Leopoldo, 179
Nachman, R. G., 88n, 263
Napoleonic restoration, 27, 30
Napoleonic wars, 43, 47
National Institute of Medical Research, U.S., 176
National Research Council. *See* Conselho Nacional de Pesquisas (CNPq)
nationalism, 91–94
Natterer, Johan von, 53
natural sciences, 8, 27, 28, 37n, 39–40, 50, 59, 61, 67, 74, 126, 135, 194
Nau, Henry, 235n, 246n, 263
naval industry, 240; in Portugal, 33; in Brazil, 240
Needell, Jeffrey D., 74n, 263
Neisser, Klauss A., 100n
Neiva, Artur, 86, 89, 90, 99, 144, 145, 146, 147, 174, 189, 263
Neri, Philip, 37n
Netherlands, the, 18, 49–50, 228
Neto, Ladislau, 54, 148
Neves, Aurora, 103n
Neves, Guilherme Bastos Pereira das (Admiral), 159
Newton, Isaac, 25–28, 32, 39n, 75
Nielsen, Waldemar A., 193n, 264
Niemeyer, Oscar, 124, 210, 212
Nobile, Umberto, 125
noneuclidian geometry, 75
Novais, Fernando A., 18, 264
nuclear research, 204, 217, 218, 226, 232–33, 236, 239, 241
Nunes, Márcia Bandeira de Melo, x n, 97n, 227n, 231, 254, 264

Oberakcker, Carlos, 18n, 264
Observatório Nacional (former Observatório Imperial), Rio de Janeiro, 78–79, 81, 162n, 169, 172

Occhialini, Giuseppe, 157, 158, 161
Oken, Lorenz, 28, 31
Oliveira, Adosindo Magalhães de, 92n, 93n
Oliveira, Álvaro Joaquim, 149n
Oliveira, Armando Sales de, 127, 128, 131n, 135, 136, 155n
Oliveira, Avelino Inácio de, 92n
Oliveira, Clodomiro de, 95n
Oliveira, Ernesto Luís de, Jr., 134, 206n
Oliveira, Eusébio Paulo de, 92, 95n, 179
Oliveira, Francisco P., 91n
Oliveira, João Batista A., 238n, 254, 264
Oliveira, Lúcia Lippi, 128n, 264
Oliveira, Neide S., 15n, 264
Ombredonne, Professor, 126n
Onorato, Ettore, 134
Oppenheim, Victor, 96
Organization of American States, 227, 228
Osório, Alberto, 110n
Ottoni, Virgílio, 86n
Oxford University, 24, 30, 100n

Pacciti, Técio, 254
Pacheco, Genésio, 144n
Paim, Antônio, x n, 45n, 80n, 109n, 110n, 119n, 120n, 206n, 264
Paraense, Wladimir Lobato, 176, 189, 254
Paraguay, 7n, 75
Pascal, 24
Passos, Pereira, 87
Pasteur Institute, 84n, 86
Pasteur, Louis, 84
Paterson, John Ligertwood, 65
Pauli, Wolfgang, 156n
Paulinyi, Erno, 222n, 264
Pavan, Crodowaldo, 143n, 177, 178n
Pedro I, King (also Pedro IV), 47, 48
Pedro II, King, 43, 48, 54, 55, 61, 238
Peixoto, Maurício Mattos, 254
Pena, Afonso, Jr., 121n
Pena, M. Valéria, 87n, 128n, 194n, 264
Pena, Osvino, 192
Penha, Adolfo Martins, 99, 145, 146n, 171, 252
Penna, Maria Luiza, 106n, 264
Pereira, Dulcídio, 162, 179
Pereira, J. Veríssimo da Costa, 91n, 92n, 264
Pereira, Jesus Soares, 92, 93n, 95, 264
Pereira, José Saturnino da Costa, 57n

Pereira, Lafaiete Rodrigues, 147
Pereira, Lino Sá, 110n
Pereira, Olinto Vieira, 103n
Pereira, Vera Maria C., 194, 264
peripheral science, 6–8, 15, 17, 211
Perlingero, Carlos Augusto, 227
Perrin, Jean, 190
Perroux, François, 135
Peru, 191
Peryassa, Antônio, 89n
Petrobrás, 97n, 227n
petroleum, 95–97
pharmaceutical schools, Brazil: in Minas Gerais, Escola de Farmácia de Ouro Preto (Ouro Preto School of Pharmacy), 102n; in Rio de Janeiro, Faculdade de Farmácia, 154n; in São Paulo, Escola Livre de Farmácia, 76; Faculdade de Farmácia, Universidade de São Paulo, 174
pharmacology, 16, 39n, 209
pharmacy, 64, 65, 133n, 152
Philosophical College, England, 26
philosophy, 26, 45, 74, 112, 117, 125, 126, 132, 134
physical anthropology, 69
physics, x n, 8, 28, 37n, 39, 45, 51, 53, 59, 60, 61, 64, 65, 66, 67, 68, 72, 75, 76, 78n, 82n, 109, 112, 125, 132, 134–35, 137, 139, 153, 155–57, 169, 170–83, 185–86, 193, 204–7, 225, 232
physiology, 56, 146, 209
Piacentini, Marcello, 124
Picaluga, I., 192n, 264
Picanço, José Correia, 63
Picard, Émile, 80
Picollo, Francesco, 134
Piéron, Henry, 109
Pieroni, R., 161
Pimenta, Aluisio, 182, 219n, 252, 264
Pinkerton, 60n
Pinto, César, 89
Pinto, Mário da Silva, 92n, 97, 168n, 169, 170, 179n, 254, 264
Pinto, O. M. de Oliveira, 264
Pinto, Ricardo Guedes Ferreira, x, 155n, 264
Pinto, Roquete, 109n, 110, 111n, 112n, 113n
Pinto Sobrinho, Ageo, 103n

Piragibe, Clélia, 235n, 265
Piratininga, Jorge Tibiriçá, 141n
Pires, João Murça, 168n
Pisa, Gabriel, 84n
Pius IX, Pope, 43n
Pizza, Salvador de Toledo, Jr., 141n
Poincaré, H., 76, 80
Poirier, Professor, 126n
Polanyi, Michael, 13, 265
Policlínica Geral, Rio de Janeiro, 86n
Pombal, Marquis. *See* Melo, Sebastião José de Carvalho e
Pombal Reform, 38–40, 44, 45
Pompéia, Paulus A., 153n, 156, 158–60, 180, 186, 206–8, 253
Portela, Pinto, 86n
Porto, Ângela, 88n, 265
Porto, Sergio, 183, 206, 207, 225, 226, 253
Portugal, 32–38, 40, 44, 50
Portuguese colonization, 7, 18, 40–49
positivism, 57, 68, 74, 75, 88, 98, 108, 149n. *See also* Compte, Auguste; Sociedade Positivista
Powell, Cecil, 161
Prado, Antonio de Almeida, 191n, 265
Prado, Caio, Jr., 18n, 47n, 265
Prado, Leal, 103n, 148n, 265
Prado, Luís Cintra do, 134
Prebisch, Raúl, 201
Prévost. *See* Chapot-Prévost
Price, Derek J. de Solla, 265
Princeton University, 157, 180
professions, 12, 17, 29–31, 133
Proto-Medicato, 63
Proudhon, Pierre J., 26
provincial committees of public teaching, Brazil, 58
Prowasek, Stanilas von, 90
Ptolemy, 24, 33
public health, 63, 68–69, 87–88, 191
Putmans, Arsène, 141n
Pyenson, Lewis, 7n, 172n, 265

Queiroz, Luiz Vicente de Souza, 141
Quental, Bartolomeu do (Friar), 37n

Rabelo, Eduardo, 86
Rabim, Júlio, 156
Rahman, A., 67, 265
Ramos, Francisco Ferreira, 82

Index

Ramos, Teodoro Augusto, 81, 82–83, 111, 131, 134, 155
Rao, Vicente, 131n
Rathburn, Richard, 91n
Rawitscher, Felix, 120, 134, 168n, 187
Readers, George, 135
Reagan, Ronald, 235
Real Horto (later Jardim Botânico), 51, 52
Real Museu, Portugal, 50
Reis, Elisa P., 40n, 73n, 265
Reis, Felipe dos Santos, 81
Reis, José, 99, 144n, 145, 146n, 171, 175, 203, 253, 265
renaissance, 6, 9, 14, 23–24, 32
Resende, Beatriz, x n
revolution of 1930 (Brazil), 105, 108, 129
revolution of 1932 in Brazil. *See* Constitutionalist Revolution
Rey, Abel, 81n
Rezende, C. Barbosa, 88n, 256
Rezende, Sérgio M., 253
Rheinboldt, Heinrich, 51, 134, 148n, 149, 151, 153n, 154, 163, 265
Ribeiro, A. C. Torres, 192n, 264
Ribeiro, Darcy, 210–12, 253
Ribeiro, Joaquim Costa, 60n, 79–80, 120, 126n, 146, 162, 163, 174, 265
Riedel, Ludwig, 53
Riemann, Georg Bernhard, 75
Ringer, Fritz K., 31n, 265
Ripper, José Ellis, 206, 225, 226
Rizini, Carlos Toledo, 168n
Rocha, Fleury da, 92n, 95
Rocha, Ismael da, 86
Rocha, Plínio Sussekind da, 163
Rockefeller Foundation, 102n, 137, 142–43, 147, 161, 169, 171–72, 178, 181, 190–94, 196, 210. *See also* medicine, tropical medicine
Rodrigues, E., 163
Rodigues, Nina, 69
Roentgen, Wilhem Conrad, 82
Rokkan, S., 259
Romani, Jacqueline Pitanguy, 205n, 265
Romeu, Antonio Soares, 134
Rondelli, Constantino, 82n
Roquete, Rubem de Carvalho, 94n
Rosa, Aldo Vieira da (Brigadier), 208, 226
Rosa, J. N. Santa, 93n, 265
Rosa, Pinguelli, 230n, 254

Rosemberg, Hans, 31n, 265
Rosenfeld, Anatol, 148n
Rosenfeld, Gastão, 203
Rothblatt, Sheldon, 46n, 265
Rothe, Otto, 120, 149, 150
Rousseau, Jean-Jacques, 26
Royal Engineering Corps, Brazil, 57
Royal Society, England, 24, 26
Rushing, Francis W., 265
Rutherford, Lord, 155, 156n, 157

Sá, Paulo Accioly de, 94n
Sábato, Ernesto, 7n, 266
Sacramento, Leandro do (Friar), 52
Sagasti, Francisco, 7, 266
Sagres School, 33
Saint-Etienne School of Mines, 62
Saint-Hilaire, Geoffroy, 50
Saint-Simon, Comte de (Claude-Henri de Rouvroy), 26
Sala, Oscar, 160, 182, 253
Salem, Tânia, 107n, 266
Sales, Dagoberto, 233n, 266
Sales, José Batista Veiga, 147
Salmeron, Roberto, 182, 204, 212, 226, 253
Salmon, George, 80
Salomon, Jean-Jacques, 13–14, 266
Salzano, Francisco M., 184, 253
Sampaio, Trajano, 141n
Sanches, Francisco, 34
Santa Casa da Misericórdia, Rio de Janeiro, 101n
Santoro, Cláudio, 212
Santos, Anísio dos, 206
Santos, Lycurgo Filho, 63n, 266
Santos, Marcelo Damy de Souza, 156, 157–59, 160, 161, 182, 185, 186, 225, 226, 250
Santos, Ribeiro dos, 39n
Saraiva, Antonio José, 33n, 34, 36, 266
Sarney, José, 241
Sauvre, Soulier de, 78n
Sawaya, Paulo, 134, 203, 253
Schaeffer, Alfred, 103n, 148–49, 150
Schenberg, Mário, 156–57, 182, 186, 253
Schirm, E., 150
Schrödinger, E., 125
Schutzer, Walter, 156

Schwartzman, Simon, x n, 4n, 40n, 67n, 73n, 93n, 97n, 117n, 120n, 122–26n, 168n, 185n, 195n, 201n, 214n, 217–19n, 227n, 231n, 235n, 242n, 254n, 264, 266, 267
science and education schools: in Rio de Janeiro, Faculdade de Ciências at the Universidade do Distrito Federal, 81, 118, 120, 121, 124; Faculdade de Educação, Ciências e Letras, 116, 117, 118; Faculdade Nacional de Filosofia, Ciências e Letras (later, Faculdade de Filosofia, Ciências e Letras at the Universidade do Brasi and, later, at the Universidade Federal do Rio de Janeiro), 118, 122, 123, 124, 125, 126, 133, 163, 180, 182, 197, 209, 225; Instituto de Educação, 121; in São Paulo, Faculdade de Educação, 131n; Faculdade de Filosofia, Ciências e Letras at USP, 82n, 120, 124, 125n, 127n, 128, 129, 130, 131, 133–34, 136–37, 143, 151–53, 155, 157, 159, 160, 161, 174, 177, 178, 179, 180, 182, 185, 186, 193, 197, 202, 237, 238
Science Citation Index, 240
scientific community, 2, 5–6, 8–10, 12–17
scientific ideologies, 4, 10–11, 25–28, 202–3
scientific societies, 16, 24, 26, 31, 44, 45, 51, 57, 202–3, 235, 267
scientism, 11, 29n
Scottish universities, 29
Secretaria de Tecnologia Industrial (STI), 225
Secretaria Especial de Informática (SEI), 234, 235, 236, 241, 267
Segre, Beniamino, 125
Seixas, Joaquim Correia de, 94n
Sellow, Friedrich, 53
Sena, J. C. da Costa, 109n
Senise, Pascoal A., 151, 182, 253
Sérgio, Antônio, 32n, 39n, 267
Serviço de Febre Amarela, 192
Serviço de Malária do Nordeste, 192
Serviço Especial de Grandes Endemias, 147, 188, 192
Serviço Geológico e Mineralógico (former Comissão Geológica do Império / Imperial Geological Commission) (later Departamento Nacional da Produção Mineral), 91–93, 95, 96, 179
Shaplen, Robert, 190, 267
Shaw, Paul Vanorden, 135
Shils, Edward, 267
Silva, Álvaro Alberto da Mota e (Admiral), 110, 204
Silva, Francisco A., 82n, 262
Silva, Heitor Lira da, 110n
Silva, José Bonifácio de Andrada e, 44, 51, 238
Silva, Maria Beatriz Nizza, 40n, 267
Silva, Martin Francisco de Andrada e, 51
Silva, Maurício Rocha e, 99, 135, 146n, 148n, 176, 179n, 203, 209n, 253, 267
Silva, Pirajá da, 99
Silva, Walzi A. Sampaio da, xi
Simonsen, Roberto, 18n, 47n, 127, 128, 267
Sinclair computers, 234
Siqueira, João Bosco de, 206
Skidmore, Thomas E., 19n, 73n, 267
Slotta, Karl Heinrich, 100n
Smilie, Wilson, 191
Smith, Adam, 27
Smith, Richard H., 206, 207
Smith, T. Lynn, 267
Soares, Glaucio Ary Dillon, 223n, 258
social sciences and humanities, 9, 75, 112, 126, 132–35, 193–94, 214, 224
Sociedade Auxiliadora da Indústria Nacional, 53
Sociedade Brasileira de Computação, 235
Sociedade Brasileira de Física, 267
Sociedade Brasileira para o Progresso da Ciência (SBPC), 202–3
Sociedade Científica do Rio de Janeiro, 51
Sociedade Positivista, 149n. *See also* positivism
Soper, Fred, 192n
Sorbonne, 130, 175n, 239
South Korea, 246
Southern Cone countries, 4n, 216, 217
Souza, A. Candido de Melo e, 267
Souza, Aníbal Pinto de, 94n, 162n
Souza, Geraldo Horácio de Paula e, 82, 191
Souza, Heitor Gurgulino, 206
Souza, José Vitorino dos Santos e, 57
Souza, Nadja V. X., x, 97n, 227n, 231, 254, 264

Index   285

Soviet Union, 155n, 228
Spencerianism, 75
Spiegel-Rösing, Ina, 267
Stammreich, Hans, 149
Standard Oil Company of New Jersey, 190
Stanford University, 154n
Stein, Stanley, 47n, 267
Stepan, Nancy, xi, 69n, 83n, 85n, 267
Stettiner, Herbert, 135, 149, 151
Stols, Eddy, 77n, 267
Strauss, Fritz, 151
Stuttgart Polytechnic Institute, 148
Suárez, Francisco, 34, 35
Surgeon-General of the Army, Brazil, 63
Switzerland, 84n
Syntex Corporation, 154n
Szyska, Gerhard, 100n

Taiwan, 246
Tandy computers, 234
Taunay, Afonso d'Escragnole, 82n, 134
Tavares, Armando Dias, 163
Távora, Juarez, 94
Technical Institute of Karlsruhe, 151n
Technical Institute, Stuttgart, 162, 173
technological research, 93, 139, 245
Teckolt, Theodore, 148
Teixeira, Anísio, 106, 118, 119–20, 121–22, 210, 212, 219
Teixeira, Glycon de Paiva, 92n, 169
Teles, Adalberto Queirós, 144
Teles, Afonso da Silva, 227
Teles, Fonseca, 131n, 158n
Thompson, J. J., 157
Tigre, Paulo Bastos, 235n, 267
Tiller, Frank, 227
Tiomno, Jaime, 180, 204, 253
Tobias, J. Antonio, 113n, 267
Todaro, M. Patrice, 107n, 267
Toledo, Paulo Saraiva de, 163
Torres, Margarinos, 192
Travassos, Lauro, 120, 188, 208n
Trompowsky (Ministry of Aeronautics), 207
Tronchon, Henri, 121
tropical medicine: Argentina, 99; University of Hamburg, 145n. *See also* Instituto Manguinhos; Rockefeller Foundation
Tupinambá, Aurélio (general), 206

Ubisch, Gertrud von, 100n
Ungaretti, Giuseppe, 135
United Kingdom, 20, 24–26, 29–31, 43, 49, 52, 74, 100n, 130, 137, 142, 155–58, 161, 173, 177, 180, 182, 183, 186, 228
United Nations Economic Commission for Latin America, 201, 216, 217
United States Agency for International Development (USAID), 219, 228
United States of America, xi, 6, 13, 31–32, 87, 91n, 95–97, 99, 119, 137, 141–42, 147, 151, 152, 154n, 158, 161, 169, 171–72, 176–78, 180–84, 190–94, 196, 204, 206, 210, 219, 225–28, 234–36
Universidade Católica, Rio de Janeiro, 126, 227n, 233, 234
Universidade da Bahia, 193
Universidade de Brasília, 208, 210–12, 219, 223
Universidade de Coimbra, 35, 39, 42, 44, 64; Colégio das Artes, 35–37; Faculdade de Filosofia, 44; Faculdade de Leis, 44
Universidade de Évora, 35
Universidade de Minas Gerais, 102, 104, 111, 182, 183, 212, 219
Universidade de Pernambuco, 182
Universidade de São Carlos, 163, 243
Universidade de São Paulo, xi, 2n, 10n, 74, 77n, 81, 100, 118, 120, 125, 127, 128, 131, 133, 136–37, 139, 142, 143, 149, 151, 153n, 155, 160, 161, 162, 164, 177, 179, 181–84, 204, 208, 216, 221, 223, 238, 240
Universidade do Brasil, x, 122, 123–24, 174
Universidade do Distrito Federal, 81, 118–22, 123, 124, 126n, 163, 174, 182, 219
Universidade do Rio de Janeiro, 113–14, 116, 121, 123, 124, 131, 154n, 204
Universidade do Rio Grande do Sul, 183, 184, 221
Universidade Estadual de Campinas, xn, 162, 212, 216, 223, 226–27, 233, 240
Universidade Estadual Júlio de Mesquita, 240
Universidade Federal de Viçosa, Minas Gerais, 103, 221
Universidade Federal do Rio de Janeiro, 97n, 216, 221, 223, 227, 229, 240
Université de l'Etat de Gand, Belgium, 77n

Université de Paris, 24, 176, 226n
Universities: of Berlin, 31, 142, 149, 151; of Bern, 84n; of Breslau, 142, 173; of Bristol, 161; of California at Berkeley, Center for Studies in Higher Education, xi; of Chicago, 182; of Giessen, 140n; of Göttingen, 151, 204; of Heidelberg, 173; of Houston, 227; of Jena, 145; of Michigan, 182; of Munich, 103n, 142, 148; of Pavia, Italy, 173; of Prague, 149; of Rostock, 148; of Southern California, 226; of Strasbourg, 151n; of Turin, 82n, 155n, 162, 173; of Vienna, 142, 173; of Wisconsin, 182; of Zurich, 148. *See also* Vanderbilt University; Yale University; etc.
Unna, J., 84n
Urca process, 157
URENCO consortium, 233

Valadares, Benedito, 103
Vale, José Ribeiro do, 100–101, 128n, 176, 203, 253, 268
Vanderbilt University, 227
Vanderley, Luis Adolfo, 82n
Vanzolini, Paulo Emílio, 16, 184, 253
Vargas, Getúlio, 72–74, 97n, 103, 106, 108, 114, 118, 119, 125, 127, 190, 195, 201, 204
Vargas, José Israel, 153n, 183, 206, 253, 254
Vasconcelos, Helena Araúlo Leite de, xi
Vasconcelos, Henrique Figueiredo, 89
Vaz, Zeferino, 176, 208–10, 212, 217, 218, 223, 225, 226, 253
Velho, Otávio G., 40n, 268
Velloso, João Paulo dos Reis, 4n, 217, 218, 226n, 238, 241, 268
Venâncio, A. Filho, 57n, 58n, 268
Venâncio, Francisco Filho, 106
Venezuela, 191
Venturi, Atílio, 135
Verney, Luís Antonio, 37–38, 268
Vessuri, Hebe, 7n, 268
veterinary, 76, 145. *See also* agriculture and animal husbandry, Ministério da Agricultura
Viana, Gaspar, 90

Viana, J. Baeta, 102, 103n, 121, 147, 178
Viana, Marcos, 225
Vidal, João Batista, 225
Vieira, Borges, 191
viradeira (turnabout), 40
Visconde do Rio Branco, 60, 61
Viscount of Sinimbu (prime minister), 58
Vital, Dom (Bishop of Olinda), 43n
Vizioli, José, 141n
Vogel, Maria Beatriz de Pena, x n
Volta, Alessandro, 28, 75

Wade, Nicholas, 246n, 268
Wallis, John, 26
Walter, Leon, 128
Washburne, Chester, 96n
Wataghin, Gleb, x n, 125, 134, 156, 157, 158, 160, 161, 162, 163, 172, 173, 185, 186, 206, 207, 225, 254
Weber, Max, 6n, 12, 247, 268
Wedekind, W., 151n
Werneck, Hugo, 102
Werner, Abraham, 44, 52
Western Europe, scientific development in, 6, 9, 14
Whitaker, A. P., 268
White, I. C., 91n
Whitley, Richard, xi, 14n, 261
Wilberg, Norman, 158
Wilhems, Emílio, 210n
Wilson, Bruce, 192n
Windaus, Adolf, 151
Wirth, John D., 19n, 73n, 93n, 268
Wittrock, Björn, 13n, 260, 268
Wladislaw, Blanka, 168n, 182, 254
women in science in Brazil, 2n
Woolgar, S., 14n
Wücherer, Otto, 65

Yale University, 102n
Yerkes Astronomic Observatory, 157
Yugoslavia, 97

Zeiss Corporation, Germany, 79n
Zocher, Hans, 96, 149, 150n
zoology, 16, 28, 50–51, 53n, 54, 88, 132, 134, 178, 187, 208n

www.ingramcontent.com/pod-product-compliance
Lightning Source LLC
Chambersburg PA
CBHW031546300426
44111CB00006BA/195